Grow Small, Think Beautiful

Grow Small, Think Beautiful

Ideas for a Sustainable World from Schumacher College

Edited by Stephan Harding

Floris Books

First published by Floris Books in 2011

British Library CIP Data available
ISBN 987-086315-835-3
Printed in Great Britain
by CPI Group (UK) Ltd, Croydon

In Memoriam

This book is dedicated to the memory of Professor Brian Goodwin (1931–2009) who so generously shared his profound insight, wisdom and friendship with everyone at Schumacher College from 1996 to 2009.

Contents

Acknowledgments

This book is very much a child of Schumacher College, and so my first thanks go to the Dartington Hall Trust which created the college just over twenty years ago. With the generous support of the Trust the college has been able to provide many people from all over the world with wonderful opportunities for deep, transformative learning experiences that have strengthened their service to society and to the Earth. In particular I would like to thank Philip Franses, my colleague on the college's MSc in Holistic Science, for his friendship and intellectual support on the journey towards wholeness that we share with our students.

I would also like to thank the editors of *Resurgence* magazine for allowing to me to include my compilation of articles written for them by James Lovelock. Thanks are also due to Guillem Ferrer in Majorca for his interview with Satish Kumar, an edited version of which appears here as one of the chapters, and to Craig Holdrege, Ezio Manzini and Richard Heinberg for permission to edit and publish chapters they hold in copyright. Ed and Liz Posey of the Gaia Foundation kindly gave me permission to include an edited version of their booklet *Gaia,* written for them by Jules Cashford.

A major thanks also to all the authors who wrote for the book. You have endured and survived my pestering and delivered manuscripts rich in meaning and insight. A final thanks to Christopher Moore and the team at Floris Books, all of whom have been a delight to work with.

Stephan Harding

Author biographies

Mark Burton is an engineer by training. He is a graduate of the MSc in Holistic Science at Schumacher College, and is currently doing his PhD at the University of Bristol on alternative banking and currency systems.

Fritjof Capra, Ph.D., physicist and systems theorist (www.fritjofcapra.net), is a founding director of the Center for Ecoliteracy in Berkeley. He is the author of several international bestsellers, including *The Web of Life* and *The Hidden Connections.* His most recent book, *The Science of Leonardo,* was published in paperback by Anchor Books in December 2008.

Jules Cashford read philosophy at St Andrew's University and did post-graduate research in literature at Cambridge, on a Carnegie Fellowship. She is a Jungian analyst and has authored many books including *The Myth of the Goddess: Evolution of an Image* (with Anne Baring) and *The Moon: Myth and Image* (Cassell Illustrated, 2003).

Shaun Chamberlin co-founded Transition Town Kingston and authored the transition movement's second book, *The Transition Timeline.* He has also served as an advisor to the UK Department of Energy and Climate Change and an academic peer reviewer for the Climate Policy journal, edited booklets on nuclear power and carbon rationing, and co-authored the All Party Parliamentary Group on Peak Oil's 2011 report into Tradable Energy Quotas.

Richard Douthwaite worked as a government economist in the West Indies before moving to Ireland to become an economist and writer with a special interest in climate and energy issues and local economic development. He is co-founder of the Foundation for the Economics of Sustainability (FEASTA) and is a council member of Comhar, the Irish government's national sustainability council and a Fellow of the Post Carbon Institute.

Brian Goodwin studied biology at McGill University and then emigrated to the UK under a Rhodes Scholarship, studying mathematics at Oxford. He received his PhD at the University of Edinburgh and then moved to Sussex University until 1983 when he became a full professor in biology at the Open University. After his retirement he joined the faculty at Schumacher College in Devon, UK, where he was instrumental in starting the MSc in Holistic Science.

Per Ingvar Haukeland is a Norwegian researcher of rural studies at Telemark Research Institute and a teacher of ecophilosophy and deep ecology at the University College of Telemark. He has an MA from University of Oregon and a PhD in ecophilosophy, education and sustainable rural entrepreneurship from University of California at Berkeley. He has collaborated with Arne Naess over several years, e.g. the book *Life's Philosophy* and *Deep Joy: Into the depths of deep ecology.*

Richard Heinberg is the author of ten books, including *The End of Growth.* He is widely regarded as one of the world's most effective communicators of the urgent need to transition away from fossil fuels. He is currently Senior Fellow-in-Residence at Post Carbon Institute, a non-profit organisation dedicated to building more resilient, sustainable, and equitable communities.

Helena Norberg-Hodge is the founder and director of the International Society for Ecology and Culture (ISEC) and its predecessor, the Ladakh Project. She is the author of *Ancient Futures: Learning from Ladakh* (1992) and co-author of *Bringing the Food Economy Home.* Norberg-Hodge's groundbreaking work in the Himalayan region of Ladakh is internationally recognised, and earned her the Right Livelihood Award.

Craig Holdrege is the founder and director of The Nature Institute in Ghent, NY, which is dedicated to research and educational activities applying phenomenological, holistic methods, including Goethean science (natureinstitute.org). He is the author of books, monographs, and articles including *Beyond Biotechnology* and *The Giraffe's Long Neck.*

Gideon Kossoff is a social ecologist/social theorist whose research focuses on the relationships between humans and the natural environment and humans and the built/designed world as the foundation for a sustainable society. Gideon served as programme administrator from 1988 until 2007 for Schumacher's MSc in Holistic Science.

Terry Irwin is the Head of the School of Design at Carnegie Mellon University in Pittsburgh, Pennsylvania. She was trained as a communication design and has Master's degrees in design from the Allgemeine Kunstgewerbeschule in Basel, Switzerland, and a Master's in Holistic Science from Schumacher College. She lectures widely on the topics of sustainable design and living systems.

Satish Kumar renounced the world and joined the wandering brotherhood of Jain monks when he was nine years old. As a young man he undertook an eight thousand mile peace pilgrimage, walking from India to America without any money. In 1973, he settled in England, taking an Editorship of *Resurgence* magazine. He is the founder of the Small School in Hartland and a co-founder of Schumacher College. He is author of several books, including *No Destination,* and *You Are, Therefore I Am – A Declaration of Dependence.*

James Lovelock is the author of more than two hundred scientific papers and the originator of the Gaia Hypothesis (now Gaia Theory) about which he has written three books, most recently *The Vanishing Face of Gaia.* A Fellow of the Royal Society, in 2003 he was made a Companion of Honour by Her Majesty the Queen. He has received numerous prizes and awards for his contributions to science, including the Blue Planet Prize, the Edinburgh Medal and the Wollaston Medal.

Ezio Manzini has been working in the field of design for sustainability for more than two decades. Most recently, his interests have focused on social innovation. He started and coordinates, DESIS: an international network on design for social innovation and sustainability. At the Politecnico di Milano, he coordinated the Unit of Research DIS, the Doctorate in Design He is the author of several design books, including *The Material of Invention, Artefacts: Towards a New Ecology of the Artificial Environment.*

Esther Maughan McLachlan is a marketing and communications specialist with particular expertise in engaging on the environmental and social issues facing business.

Sergio Maraschin is a field geologist who was a senior environmental scientist for Shell before coming to Schumacher College where he studied for his MSc in Holistic Science. He currently lives in Portugal where he is creating a holistic learning centre based on permaculture principles.

Peter Reason retired in 2009 from an academic career at the University of Bath in which his primary contribution was in the theory and practice of participatory action research. He was Director of the Centre for Action Research in Professional Practice (CARPP) and co-founder of the MSc in Responsibility and Business Practice. In these programmes he pioneered graduate education based on collaborative, experiential and action-oriented forms of inquiry.

Julie Richardson taught ecological economics and international development at the Universities of London and Sussex and has worked in sustainable development for over twenty years in Africa, Asia, Latin America and Europe. She is a graduate of the Master's in Holistic Science at Schumacher College, where she is currently Head of Economics and is co-author of *The Triple Bottom Line: Does It All Add Up?*

Rupert Sheldrake is a biologist and author of more than eighty technical papers and ten books, including *A New Science of Life*. He was a Fellow of Clare College, Cambridge, and Director of Studies in cell biology, and also a Research Fellow of the Royal Society. From 2005–2010 he was the Director of the Perrott-Warrick Project, funded from Trinity College, Cambridge. He is currently a Fellow of the Institute of Noetic Sciences, near San Francisco, and lives in London.

Stephen Sterling is Professor of Sustainability Education, Centre for Sustainable Futures (CSF), at the University of Plymouth, UK. His research interest is in ecological thinking, systemic change, and learning at individual and institutional scales to help meet the challenge of accelerating the educational response to the sustainability agenda. His last book, with P. Jones and D. Selby, is *Sustainability Education: Perspectives and Practice across Higher Education*.

Nigel Topping is a mountaineer, industrialist and social entrepreneur. He spent nearly twenty years in industry transforming old businesses using radical approaches to employee relations and 'lean manufacturing' thinking. Nigel is currently Chief Innovation Officer at the Carbon Disclosure Project, a global NGO working to redirect the flow of global capital towards climate change solutions by using the power of information.

Colin Tudge writes about a wide range of topics, including evolutionary biology, religion, politics, food and farming. His recent books include *The Secret Life of Trees* and *Feeding People is Easy*. For further details about Colin's work on food and farming, see the websites of the Campaign for Real Farming and the College for Enlightened Agriculture.

Introduction: The Background to This Book: Schumacher College and Brian Goodwin

STEPHAN HARDING

Schumacher College is a world-renowned centre for transformational learning. For twenty years the College has been inspiring people from all over the planet and from all walks of life to explore the roots of the current global crisis and to actively engage in its resolution out in the wider world. The College is a daring and visionary initiative of the Dartington Hall Trust, whose founders, Dorothy and Leonard Elmhirst, began a radical experiment in rural regeneration in the 1920's here on the Dartington Hall estate. The inspiration for this ground-breaking project came from their mentor, the great Indian writer and philosopher, Rabindranath Tagore.

It was my extreme good fortune to have joined the College as its Ecologist in Residence in October 1990, just three months before the launch of its very first course – a five week *tour de force* on Gaia theory led by James Lovelock. Housed in the Old Postern, a lovely medieval building nestled amongst the extensive woods and fields of the beautiful Dartington Hall estate, the College has gone from strength to strength since that first epoch-making exploration of the science, life and soul of our breathing planet.

One of the founders and great inspirational figures at the College since the very outset has been the Indian sage, writer and activist, Satish Kumar, who over the last two decades has invited the most important ecological thinkers and activists from around the world to teach at the College. After James Lovelock came Helena Norberg-Hodge, and soon after we were graced with the presences of luminaries such as Fritjof Capra, Rupert Sheldrake, Vandana Shiva, Hazel Henderson, Jonathon Porritt and Arne Naess, to name just a few. Year after year, great minds and souls have come to the College as teachers, participants, helpers and facilitators. Inspired by the Indian ashram tradition we have learnt together, cooked, cleaned and gardened together and lived together with a deep sense of community.

One of these great people was the eminent mathematician and biologist, Brian Goodwin, Professor of biology at the Open University. Brian first came to the College in 1996, invited by Vandana Shiva to share his critique of neo-Darwinism with participants on her short course. Brian's visit was a defining moment for the College. In his brilliant session he explained how Darwin's social context (rapacious Victorian capitalism) had influenced his concept of natural selection, and also how natural selection on its own does not in any way explain the deeply mysterious origins of biological order and form. Brian's intellect, his clear delivery, his humility and his humour were deeply impressive, and so we were delighted when, a few months later, close to his retirement from the Open University, he asked whether he could join the College faculty. He needed no salary – all he wanted was to live in Deer Park Cottage, a small house set in its own woodland on a remote part of the Dartington estate. Once he was happily settled into his beloved cottage, Brian, myself and Anne Philips (the Director of the College at the time) began the task of establishing the world's first MSc in Holistic Science in partnership with the nearby University of Plymouth.

It had long been Brian's dream to establish a Masters degree where students could explore a participatory science of qualities, values and interactions that underpins an ecological worldview, and where they could explore new transdisciplinary methodologies that go beyond reductionism in understanding whole systems. The programme began in 1998 and is now in its thirteenth year with around one hundred graduates all over the world. The success of the MSc has made it possible for College to launch a new MA in Economics for Transition, led by Julie Richardson, one of our MSc graduates, and Jonathan Dawson. This degree, also offered in partnership with the University of Plymouth, begins in September 2011.

For thirteen years I was Brian's closest colleague, sharing ideas, visions, insights and revelations with him and with our students. I could not have had a better teacher and mentor. He gave me what I so desperately missed during my scientific training at Durham and Oxford: a profound understanding of the limitations of the 'selfish gene' view of life, and a sense of how one could bring soul into science. Just before Brian passed away in the summer of 2009, he asked me whether I would help him realise another of his dreams: the editing of the book that you are now holding in your hands – the first 'Schumacher College Reader' on solutions to the global crisis. It has given me great delight and a deep sense of connection with Brian to fulfil one of his last wishes, which has been made possible by a timely commission from Floris Books.

Each chapter has been authored in Brian's honour by a Schumacher College teacher who met and worked alongside him at the College. To begin with, I wanted to organise the book into subject areas, but in the end this approach was only partially successful. In Brian's playful spirit of emergence, I let the chapters decide for themselves where each of them wanted to be within the book until the wholeness of the text appeared by itself.

We begin with the importance of education for solving the global crisis, and then delve into the contributions that philosophy, spirituality and mythology can make towards the same end. It is then the turn of science, followed by development, Transition thinking, economics, energy provisioning and business. Then design offers some solutions before we finish where we started – with education.

Some of the authors are very big names, others not, but all have important ideas to offer in our search for ways out of the crisis that are socially just and ecologically sound. You will find a wide range of approaches here, each representative of the work of Schumacher College, each influenced, to a larger or lesser degree, by Brian Goodwin, to whose memory this book is dedicated.

Stephan Harding,
Schumacher College,
February 2011

1. Towards Anticipative Education – Learning by Design

STEPHEN STERLING

Suppose that the goal of an education system is for people to work cooperatively in community while exploring their individual potential for creative participation in developing and maintaining a sustainable relationship with the natural world.
What would it look like?

BRIAN GOODWIN (2007, p.337)

This rich quote from a paper by Brian Goodwin reflects his deeply holistic view of the world – in this case, in relation to the role, scope and purpose of education. What is remarkable about it is just how much meaning Brian managed get into one sentence. In this chapter, I outline some thoughts about wholeness and education, which attempt to help answer his question regarding the nature of such education. As Brian knew, re-thinking education so that it is fit for our times can no longer be an interesting intellectual exercise: this is for real. The systemic interlocking of myriad issues related to climate change, the end of cheap energy, and financial instability against a background of increasing inequity and global population means that education as a whole needs to wake to its responsibility to help shape a liveable future. As Lester Brown states:

> we are addressing not just the future of humanity in an abstract sense, but the future of our families and our friends. No generation has faced a challenge with the complexity, scale, and urgency of the one that we face.
> (Brown 2011, p.xi)

Curiously, Brown's book doesn't mention *learning* once, but his wake-up call is not new: indeed, its familiarity is a telling indicator of our Western culture's seeming inability to learn in any deep way from the recent past. Yet our current collective ability – locally, nationally, and globally – to respond adequately to threats depends on learning. But it's important to make a distinction between two types: *anticipative learning*, or 'learning by design' on the one hand, and *reactive learning* or learning 'by default' on the other. Default learning happens when events impress themselves on the learners' consciousness, by surprise, shock or crisis. Learning by design, by contrast, implies a prior awareness, a willingness and intention to learn in response to a perceived innovation, threat or opportunity. The former is a reactive response; the latter is an anticipative response.

After nearly forty years working in environmental and sustainability education, I share Sara Parkin's frustration articulated in her latest book *The Positive Deviant – Sustainability Leadership in a Perverse World* (Parkin 2010) regarding the slowness of real social change, of anticipative learning towards sustainable pathways. But what of the education community and its role in this journey? Biologist Mary Clark suggests:

> Education can never be apolitical, 'objective' or 'value neutral': it is – and ever must be – a political endeavour. It either *moulds* the young to fit in with traditional beliefs, or it *critiques* those beliefs and helps to *create* new ones.
> (Clark 1989, p.234. Author's italics.)

Yet Clark's 'critique-create' mode depends on prior learning within the education community itself. Whilst there have numerous high level calls for the reorientation of educational systems to embrace environmental issues and sustainable development ever since the UN Conference on the Human Environment of 1972 – and a good deal of international activity has been associated with the current UN Decade of Education for Sustainable Development (DESD) – overall education remains 'a slow learner'. Nearly four decades later – and incredibly perhaps, given that all education is in some sense about preparing for the future – most education still makes little or no reference to these overarching contextual issues. Here is a profound paradox: education is held to be a key agent of change, and yet is a contributory element of the unsustainability problem it needs to address. How do we work towards transformative learning in a system that itself is intended to be the prime agency of learning? How do we accelerate its *'response-ability'*, that is, the ability of the educational community as a whole to respond adequately to the emerging conditions of threat and opportunity that face our communities, our graduates, and our children? Embedding sustainability concepts within systems that otherwise remain unchanged may be a start but will not be sufficient, because the problem is deeper. With this in mind, we need to see education systems – and therefore associated policies and practices – as a subsystem of the society and culture they serve. As I have written elsewhere:

> the policy and practice of Western and Westernised education is still largely built on the assumptions and epistemology of a previous age, rather than fully responsive to the conditions and needs of our time. If this is the case, then reliance on education as a critical path to a more secure and liveable future seems a risky strategy. Such a future will not be assured without learning: the question is whether formal education can and will be part of this learning. The answer hangs on whether the educational community – policymakers, theorists, researchers and practitioners – can itself experience some quality of transformative learning and awakening so that the education provision that in turn then evolves can be transformative rather than, as at present, conformative
> (Sterling 2009a, p.105)

In short, we are confronted with a double learning challenge. We need to make a distinction between two arenas: *designed learning* and *institutional learning*. The former is the explicit and day to day concern of all educational programmes: it is planned, resourced and provided for different groups such as school pupils, tertiary level students, and adults in community education. *Institutional learning* refers to the social and organisational learning that the policy makers and providers may themselves undergo or experience: for example, government educational departments, schools, universities, and educational agencies. In other words, sustainability requires learning *within* educational systems, not just learning *through* educational systems. My experience working in the field has proven the principle many times over: significant change in the latter requires significant change in the former. To elaborate further, in response to the crisis of unsustainability, most educators – and increasingly, politicians – will ask: 'What learning needs to take place amongst students?'. This is a perfectly valid and important question, but it begs a prior and deeper question: what changes and what learning need to take place amongst policymakers, amongst senior management, amongst teachers, lecturers, support staff, amongst parents, amongst employers, etc. so that education itself can be more transformative and appropriate to our times? The first question stays within what learning theorists call *first order change*, that is, more of the same: accommodatory change which doesn't affect the system as a whole. But what if the system itself needs changing? This invokes at least second order change which involves a re-examination of assumptions – towards a shift of consciousness, a changed intelligence which is both connective and collective. This is a deeper and systemic learning response, which needs to happen in four areas: personal, professional, organisational and – beyond formal education – in the community (social learning). The most resistant area is change in the institution and organisation – but this is being squeezed to some extent by growing awareness of individuals at one level; and shifts in social values and behaviours at the level of community and public debate at another.

My doctoral thesis on whole systems thinking and education (Sterling 2003), much of which was written at Schumacher College over a period of years – and benefited greatly from the thinking manifested there – grappled with a question that exercises me to this day: why education as a whole, and environmental and sustainability education in particular, are limited in their ability to make a positive difference to the human and environmental prospect by helping assure a more sustainable future. If I were forced to answer my own question in one sentence, it would be this. *Education is still fundamentally reductive.* Despite the discourses of postmodernism and post normal science, the rise of complexity theory, and everyday evidence of the systemic nature of the world, the fundamental building blocks of the prevalent education epistemology – reductionism, objectivism, materialism, dualism, and determinism – largely prevail, reflected from the dominant cultural worldview and exerting influence in purpose, policy and provision as well as in educational discourse.

These habits of thought might not be consciously recognised by most practising educators, but they are no less powerful. They reside in the subterranean geology of education, invisible in themselves but manifested in the educational landscape above: single and bounded disciplines, a resistance to interdisciplinarity, separate departments, abstract and bounded knowledge, belief in value-free knowing, privileging of cognitive/intellectual knowing over affective and practical knowing, a reluctance to consider ethical issues, prevalence of technical rationality, transmissive pedagogy, analysis over synthesis, and an emphasis on first order or maintenance learning which leaves basic values – of staff, students and institutions – largely unexamined. Hence, it can be argued that education shares in Scharmer's view of a 'massive institutional failure: we haven't learned to mold, bend, and transform our centuries old collective pattern of thinking, conversing, and institutionalizing to fit the realities of today' (Scharmer 2006, p.3).

In other words, the dominant educational paradigm maintains resilience even as the encompassing conditions of complexity, systemicity, uncertainty and unsustainability become ever more evident in wider society. As I've written elsewhere:

> The paradox of education is that it is seen as a preparation for the future, but it grows out of the past. In stable conditions, this socialisation and replication function of education is sufficient: in volatile conditions where there is an increasingly shared sense (as well as numerous reports indicating) that the future will not be anything like a linear extension of the past, it sets boundaries and barriers to innovation, creativity, and experimentation.
> (Sterling 2009b, p.19)

This is not the whole story of course. Over the past few decades there has been a growing education for change movement encompassing such emphases as development education, community education, peace education, human rights education, anti-racist education, humane education, futures education, environmental education, and sustainability education. Despite genuine progress, such manifestations of 'education for a better world' lie partly within and partly without the dominant modernist worldview, which still prevails in education and wider society. Richmond (2009, p.3) writing in the UNESCO mid-term review of the Decade of Education for Sustainable Development, states that a 'paradigm shift in thinking, teaching and learning for a sustainable world' needs to be realised, and an holistic approach to teaching and learning is vital and urgent. Logically, sustainability necessitates a deep questioning and learning response in educational thinking and practice as a whole, just as it does in myriad other human activities, whether economics and business, design and construction, agriculture and energy, trade and aid, health and welfare, and so on. It cannot simply be a matter of 'add-on', but is a matter of re-design with a shift of emphasis from relationships based on fragmentation, control and manipulation towards those based on participation, appreciation and self-organisation.

Instead of educational thinking and practice that tacitly assumes that the future is some kind of linear extension of the past, we need an *anticipative* education in full recognition of the new conditions and discontinuities which face present generations, let alone future ones: such as the massive challenges of global warming, species extinction, economic vulnerability, social fragmentation and migration, endemic poverty, peak oil, and more positively, the rise of localism, participative democracy, green purchasing, ethical business, and efforts to achieve a low carbon economy. The heart of such an education is an ecological orientation, in the Schumacherian sense of being founded on an holistic, systemic, participative, or living systems view of the world. This emerging cultural paradigm has the depth and rigour to support a redesigned educational paradigm that is in essence *relational*, engaged, ethically oriented, and locally and globally relevant. The real issue is less 'how do we educate for sustainability' (as important as this is at one level) but rather to pay deep attention to *education* itself – its paradigms, policies, purposes, and practices and their *adequacy* for the age we find ourselves in. An ecological view implies *putting relationship back into education and learning* – seeking synergy and coherence between all aspects of education: ethos, curriculum, pedagogy, management, procurement and resource use, architecture, and community links. The emphasis is on such values as respect, trust, participation, community, ownership, justice, participative democracy, openness, sufficiency, conservation, critical reflection, healthy emergence and a sense of meaning: an education which is sustaining of people, livelihoods and ecologies.

In writing the 2001 Schumacher Briefing *Sustainable Education* (Sterling 2001) I explored the potential of ecological and systemic thought as the basis of a unifying theory of education and learning which integrated the best of past liberal education practice with the newer emphases such as transformative learning, capacity building, creativity and adaptive management considered part of the new sustainability agenda, and suggested steps to help achieve constructive change at all levels. This is more than an isolated 'education for sustainability' programme, it is about a shift of personal consciousness and educational culture. I suggest that this involves movement in three interrelated areas of human knowing and experience: *perception* (or the affective dimension), *conception* (or the cognitive dimension), and *practice* (or the intentional dimension). In each of these areas, higher order learning towards an ecological consciousness and competence involves 'responsive movements', that is towards greater:

—'respons-ibility' – *an expanded and ethical sense of concern/ engagement (perception);*
—'co-rrespondence' – *a closer knowledge match with the real world including pattern, consequence and connectivity (conception); and*
—'respons-ability' – *the ability to design and take integrative and wise action in context (practice).*

In organisational terms, a move towards an ecological educational culture implies changes across the nesting levels of paradigm, purpose, policy and practice (see Table 1).

Paradigm	instead of education reflecting a paradigm founded on a mechanistic root metaphor and embracing reductionism, positivism, and objectivism, it begins to reflect a paradigm founded on a living systems or ecological metaphor and view of the world, embracing holism, systemisism and critical subjectivity. This gives rise to a change of ethos and purpose...
Purpose	instead of education being mostly or only as preparation for economic life, it becomes: a broader education for a sustainable society/communities; sustainable economy; sustainable ecology. This expanded sense of purpose gives rise to a shift in policy.
Policy	instead of education being viewed solely in terms of product(courses/materials/qualifications/educated people) it becomes: much more seen as a process of developing potential and capacity through life, at individual and community levels through continuous learning. This requires a change in methodology and practice...

Practice	instead of education being largely confined to instruction and transmission, it becomes: much more a participative, dynamic, active learning process based more on generating knowledge and meaning in context, and on real-world/situated problems and issues.

Table 1. Based on Sterling 2004.

All this may seem a long way from the realities of everyday institutional life. But the question is how far these realities can correspond to the global and local realities of everyday life beyond academe. It seems there is a mismatch, a need for a fundamental re-calibration. More positively, there is significant evidence – at least in my own field of higher education – of growing interest amongst academics and students in addressing sustainability, a growing realisation that it has implications right across institutional life, and, in some pockets, a willingness to experiment with new devolved management models, flexible, innovative and broader based curricula, and an attempt to develop inclusive community within and beyond the institution.

In my 2001 Schumacher Briefing, I outlined a three stage model of institutional change, making a distinction between a content led *accommodative* response, a more values oriented *reformative* response, and a deeper *transformative* response. In the period since, there has been movement towards interest in whole institutional change, interdisciplinarity, resilience and transformative learning, but few institutions can yet point to excellence in these regards. The last – transformative – stage I characterised thus:

> A *re-design* on sustainability principles, based on a realisation of the need for paradigm change. This response emphasises process and the quality of learning, which is seen as an essentially creative, reflexive and participative process. Knowing is seen as approximate, relational and often provisional, and learning is continual exploration through practice. The shift here is towards 'learning *as* change' which engages the whole person and the whole learning institution, whereby the meaning of sustainable living is continually explored and negotiated. There is a keen sense of emergence and ability to work with ambiguity and uncertainty. Space

> and time are valued, to allow creativity, imagination, and cooperative learning to flourish. Inter- and transdisciplinarity are common, there is an emphasis on real-life issues, and the boundaries between institution and community are fluid. In this dynamic state, the process of sustainable development or sustainable living is essentially one of learning, while the context of learning is essentially that of sustainability. In this way, sustainability becomes an emergent property of the sets of relationships that evolve. This response is the most difficult to achieve, particularly at institutional level, as it is most in conflict with existing structures, values and methodologies, and cannot be imposed. The descriptive term here is 'education *as* sustainability' or 'sustainable education'.
> (Sterling 2004, p.59)

A decade after I first wrote a version of this image with Schumacher College in mind, I still view it as the closest realisation and manifestation of these principles, reflected in its reputation as an inspirational centre which embodies transformative learning (Phillips 2008). Few formal and mainstream institutions can emulate Schumacher College in any thorough way but there are elements that can be and are being reflected elsewhere, and often space for small-scale but transformative learning experiences can be found. In short, educational institutions need to become less centres of transmission and delivery, and more centres of transformation and critical inquiry, less teaching organisations, more learning organisations critically engaged with real world issues in their community and region. They would be less engaged in 'retrospective education', following on from past practice, and more involved in 'anticipative education': that is, in Scharmer's words, 'learning from the future as it emerges' (Scharmer 2006, p.5).

Short of inspiring paradigm change, systems theorists know that the next most powerful lever for changing a system is to re-vision its purpose or goal (Meadows 2009). Taking Brian Goodwin's quote again, if we delete the two first words and the concluding question, it reads, '*The goal of an education system is for people to work cooperatively in community while exploring their individual potential for creative participation in developing and maintaining a sustainable relationship with the natural world*'. An inspiring way to end – or rather, to begin.

References

Brown, Lester (2011) *World on the Edge: How to Prevent Environmental and Economic Collapse*, Earth Policy Institute.

Clark, Mary (1989) *Ariadne's Thread – The Search for New Ways of Thinking*, Macmillan, Basingstoke.

Goodwin, Brian (2007) *IJISD* 'Science, spirituality and holism within higher education', *IJISD*, vol 2, nos.3/4.pp.332– 39.

Meadows, Donella (2009) *Thinking in Systems – a primer*, Earthscan, London.

Richmond, Mark (2009) 'Foreword' In Wals, Arjen, *A Review of Contexts and Structures for Education for Sustainable Development – Learning for a Sustainable World*, UNESCO, Paris.

Parkin, Sara (2010) *The Positive Deviant – Sustainability Leadership in a Perverse World*, Earthscan, London.

Phillips, Anne (2008) *Holistic Education – Learning from Schumacher College*, Green Books, Dartington.

Scharmer, Otto (2006) 'Theory U: Leading from the Future as it Emerges – the Social Technology of Presencing', Fieldnotes, September–October, The Shambhala Institute for Authentic Leadership, Halifax, NS.

Sterling, Stephen (2001) *Sustainable Education – Re-visioning learning and change*, Schumacher Briefing no.6 Schumacher Society/Green Books, Dartington.

—, (2003) *Whole Systems Thinking as a Basis for Paradigm Change in Education: Explorations in the Context of Sustainability*, (PhD thesis), Centre for Research in Education and the Environment, University of Bath. See: www.bath.ac.uk/cree/sterling.htm.

—, (2004) 'Higher Education, Sustainability and the Role of Systemic Learning', in Corcoran PB & Wals, AEJ (editors), *Higher Education and the Challenge of Sustainability: Contestation, Critique, Practice, and Promise*, Kluwer Academic, Dordrecht.

—, (2009a) 'Sustainable Education' in Gray, D., Colucci-Gray, L. and Camino, E. *Science, Society and Sustainability: Education and Empowerment for an Uncertain World*, Routledge, New York and London.

—, (2009b) Towards Sustainable Education, *Environmental Scientist* 18, no.1: 19–21.

2. Be the Change that You Seek – Wherever You Are!

Life Philosophy and Depth-ecology of Place

PER INGVAR HAUKELAND

In weighing the fate of the earth and, with it, our own fate, we stand before a mystery, and in tampering with the earth we tamper with a mystery. We are in deep ignorance. Our ignorance should dispose us to wonder, our wonder should make us humble, our humility should inspire us to reverence and caution, and our reverence and caution should lead us to act without delay to withdraw the threat we now pose to the earth and to ourselves.

JONATHAN SCHELL, *FATE OF THE EARTH* (1982)

The Norwegian philosopher Arne Naess (1912–2009), who coined the term *deep ecology*, made his final trip to his cabin Tvergastein in May, 2008. As we were breathing in the view, we spoke softly about the need to care for the wonderful world we are so lucky to be part of. He saw a need for more collaboration between the peace movement, social justice movement and the (deep) ecology movement; a deepening care for humans does not exclude a widening care for other living beings. Naess was optimistic on behalf of the twenty-second century, but his optimism depended on what you and I do today. Gandhi said: 'Be the change that you seek!' to which we can add, '...wherever you are!' Changes need to *take place* everywhere.

Life philosophy of wonder

It is winter in Breskelia, a small hamlet at the foothill of Lifjell mountain in Telemark, Norway. From the window of my study, I see directly into a forest dressed in white. A small bird comes up to the window and pokes at the glass with its beak a clear message: 'There is no food left!' As I walk out to give them more, I hear a song of spring from the treetops. The song awakens in me a feeling of the mysterious and magical unity of all life. It makes me humble and grateful to be alive.

Philosophy starts with wonder and ends with wonder. Rachel Carson, whom Naess called the mother of the deep ecology movement, writes in *The Sense of Wonder:* 'It is a wholesome and necessary thing for us to turn again to the earth and in the contemplation of her beauties to know of wonder and humility' (Carson 1965).

We need to live more integral lives, between deeper values and concrete decisions, so as to realise peaceful, just and ecological sustainable changes wherever we are. Many associate the prefix *eco* in *ecophilosophy* and *ecosophy*, perhaps too narrowly, with nature protection, but there is a need for a stronger life-orientation and an integral approach to nature, culture, community, economy and spirituality in places where we live.

The word *eco* comes from the Greek *oikos,* usually translated as *household,* but it can also be translated as *home* and interpreted as our *home in life.* In this view ecophilosophy or ecosophy can be understood as life philosophy, and ecocrisis as life crisis and ecocentrism as life-centered. A personal life philosophy is something we all need to develop, inspired by a sense of wonder and a love of wisdom into all related aspects of our home in life and of the home to all life on this wonderful planet. Such deepened and widened sense of wonder can enhance the necessary humility and reverence that makes us act in awe toward the web of life.

Life seen as an open landscape

In *Life's Philosophy* (2003), Naess and I wanted to challenge people to re-assess the direction in their life, to go deeper into the relationship between what they value and what they do. I used the following metaphor: 'To live is like traveling through a landscape with both easy and broken terrain, light and dark places, all concealing the unexpected ...to feel at home in life requires both moving toward a goal and simply being' (p.1–2). The metaphor can help us understand better the relationships between: (1) map and territory; and (2) travelling and resting.

The relationship between map and territory is like the relationship between the world of thought (mental maps) and the world of concrete things (reality). A map is not the territory, as the name is not the thing named, but it says something about its abstract structures. In the forest behind our house, there is a rock known as *Bakstekjerringa* (*Old baker woman*) in our place-based culture. Someone put up a sign in front of the rock telling three stories. The first story is about an old woman who had to climb the rock in order to escape from an angry bear. The second is about an old baker woman who refused to give some bread to a poor beggar and immediately was turned into the rock. The third story tells us that the rock is a rock that came with the ice over ten thousand years ago. Which story is the real one? If we view the life of the rock with only one map, or tell only one story, would we not make the rock less than what it is and impoverish our world of experience?

We need to be critically aware of the maps we use. Some are more dominant than others, such as the Cartesian map of mechanism, objectivism and dualism. Alfred North Whitehead warns us of 'the fallacy of misplaced concreteness', where we take something abstract as the concrete. If we think of nature as 'out there', as a machine for human purposes, and we act upon this conception while we in reality are an integral part of nature, we have not only an ecological problem, but also an existential problem.

Naess makes a distinction between *abstract structures of reality* (mapped world) and *concrete content of reality* (experienced world). Science can tell you how fast you fall from a ladder, but not how it feels to fall, nor how fast you fall when you fall in love. We have access to the concrete content of reality through direct, spontaneous experiences, which is a world of wonder and magic where everything is alive and speaks. The Norwegian poet Hans Børli tells of an early morning walk when something suddenly happens:

> A clear bellflower-morning, you may experience one of these rare moments when you, in a way, rise out of yourself and into all-life. In these moments of revelation you sense all things directly – no name, no shadowing word-image stands between you and the naked, own-life of things. Nothing is 'as something', you can erase the comparisons, analogies – the

> things are what they are under the sun, and you converse with them in a silent language of fragrance and premonition. You are within the wordless poetic sphere, where the word-poem never can follow you. But it lasts so short, it passes by like a dream, and you wake up again in a world where the impressions always strive after an 'as' – an image to be registered in the archive of experience
> (Børli 1992, p.161; my translation).

Naess would say we rise into our *ecological Self* in such spontaneous experiences. Through a process of identification, we expand our understanding of our Self (with capital S). When I was a kid, there was a beautiful fir tree in the nearby forest that I loved very much. It was the chief of all trees, and the first tree I had a deep relation to. Under its protection we built our teepee, and we loved to climb it and it loved to be climbed. After several years abroad, I went back to honour the old chief. As I came to the area, I was shocked – it had all been cut down and cleared. I searched and found the tree stump of the old chief, sat down and cried. A part of my Self had died. When we expand our Selves to encompass others, we would care for them as we would care for our Self.

We can speak of three ways of travelling in life's landscape: (1) travelling by map (thinking), (2) travelling by the territory (feeling and sensing), (3) travelling by compass (intuiting). We need to include all three to best orient ourselves. This can be illustrated with the metaphor of a sailboat. The boat is our Self. Feelings are the sail. No sail, no movement. The source of movement is the wind (spirit). Our thinking is the map we navigate after. Reason is the rudder that sets out the course, but there are two kinds of reason. A shallow reason that says something is reasonable, like building a new parking lot in the city-centre, when it is the logical thing to do given there are so many more cars and since there is a political will to do it, it is technically feasible and it is financed. A deep reason that says something is reasonable if it coincides with our deeper values in life. We related this deeper reason in *Life's Philosophy* to the term *ratio* in Spinoza's philosophy. It is also related to what Blaise Pascal (1623–62) said in *Les Pensées*: 'The heart has its reasons of which reason knows nothing.' With this deep reason a new parking lot may not be that reasonable. We need to question the reason behind the direction we individually and as a society are heading in.

Finally, it is important to rest in life's landscape, to regain our breath (spirit). To rest means different things for different people – a walk in the forest, family, friends, a concert, a good book or simply dwelling in life's many intrinsic values. You have to find out what makes you rest, and then seek that as often as you can, since it gives you the power to move.

Into the depths of deep ecology

Fritjof Capra says in the book *Web of Life* (1996) that *deep ecology* is the new paradigm he sees arise in our society. Arne Naess coined the term *deep ecology* in the early seventies to describe an approach that went deeper into the problems as well as into the solutions of the ecological crisis. People adhering to this approach did so from various total views (ecosophies), inspired by philosophy or religion, and they shared certain values and visions, for example seeing all life as having intrinsic value and the need for a non-violent revolution of how we think and act in all areas of life. Naess described also a *shallow ecology movement*, known as office-desk environmentalism, where it makes sense to protect nature in so far as it has value for humans, and where the solutions to the ecocrisis come from technological fixes.

There is a similar shallow and deep approach to the current economic crisis. Instead of going to the roots of the problems, governments have poured billions of dollars into the same institutions that in many ways are the problem. Is that a long-term solution? Why we do not change seems to have to do with a consumer-addiction to global economic systems that are themselves addicted to our addictions: 'Got to have it, just got to have it!' We are in a double-bind. On the one hand we are told to reduce our consumption to solve the ecological crisis, while on the other hand, we are told to increase our consumption to solve the economic crisis. There are alternatives, as we see in transition towns, bioregional movements and voluntary simplicity, where we consume lesser, smaller, simpler, slower and smarter, but these alternatives can only be realised, it seems, by breaking the pattern of addiction.

The cultural ecologist David Abram, the initiator and co-founder of the Alliance for Wild Ethics (AWE), together with Stephan Harding, Per Espen Stoknes and myself, uses the term *depth ecology* to avoid the dichotomy between the deep and the shallow, and to emphasise what he finds inspiring with deep ecology writings, namely how we are an integral part of nature. We have in AWE explored this in our various work (see References), especially related to the enchanted world of wonder and magic.

We can speak of four depths to depth ecology: (1) depth of immersion, which goes deeper into the soil of experience, to those deep, mystical experiences that spontaneously mark our soul and alter us in some way; (2) depth of consciousness, which entails the coming together in awareness, individually and collectively, through a process of deep inquiry into our values and visions; (3) depth of consequence, which addresses consequences of our awareness through various plans of actions in the different areas we are at, including how to overcome that which inhibits profound changes and how to promote that which enhances such changes; (4) depth of emersion, which addresses how we realise our true Self through what we do in concrete situations. There is a natural sequence to how these depths co-evolve, but they are also depths that keep on taking place in parallel and interrelated processes.

Depth ecology of place

Naess speaks warmly about Tvergastein as The Place (with capital P). He says of place:

> The loss of belonging to a place is noticeable, the longing is still there, and this emphasises the need to express what it means to belong to a place. It strengthens the tendency to develop a sense of home, to re-awaken the inner relation to oneself and the environment. This is of great importance in the deep ecology movement.
> (In Haukeland, ed. 2008; my translation.)

In my *house* (*oikos*), there are especially two *rooms* I like the most. The first is the mountain called *Bryggefjell,* that rises majestically above the treeline and demands attention from the horizon. The mountaintop is an image of getting a clear view, not just of the outer landscape but also of the inner landscape. We climb the mountain to see more of who we are. The second *room* is along the creek. In between the tree trunks, I see a small waterfall. It has formed beautiful ice sculptures last night, but the water still throws itself over the cliff, splashing into a rock and flying everywhere like sparkling diamonds in the morning sun. I often follow the drops in my imagination down the creek, into the river, lakes, canals, fjord and ocean, and from there up to the clouds, flying with the wind, and falling like as rain on Bryggefjell. What an adventure! Last spring my family and I followed the creek, to find its source, which was a wonderful small pond at the end of a marsh. The whole water-planet can be found in this small pond.

A term that can be useful in strengthening a sense of place is *genius loci*. In ancient Rome, the expression was used to denote a spirit that protected a place, but today it is used to describe the uniqueness of a place, its atmosphere. The uniqueness is in the air. The Norwegian architect Christian Norberg-Schulz (1926–2000) is known for bringing the term into architecture, inspired in part by phenomenology. It makes us aware what is *in place* and what is *out of place*.

When we moved here, there was a big, old, rusty, yellow snow plough directly opposite to our entrance. Neighbours said the plough had been there for thirty years, placed there after a small quarrel between neighbours. The guy that placed it there had been dead for over twenty years. I asked if I could remove it, which the relatives were happy for. I called the company that recycles iron, and the following day it was gone. Neighbours were so pleased that it was gone. It had become taken for granted in their everyday landscape. Now they saw how nice the place was around the plough. We can all be more attentive to what is *in place* and what is *out of place*, but sometimes we need to travel outside our place and come back after a while in order to see it with fresh eyes. We need to de-learn being a passive observer and re-learn being an active participant in shaping the place we live in, but not just there, we need to be the changes wherever we are – and we are always placed somewhere, either in workplaces, places of leisure, market places, and so on. In my work I focus on place-based education and place-based entrepreneurship as ways to enhance a peaceful, just and ecological sustainable future, and we try to do so in a *glocal* context, that is we see the global in the local and the local in the global.

We may also here speak of four depths to a depth ecology of place. First, the depth of immersion gets at how deeply we are rooted in the places we call home. Second, the depth of awareness makes us conscious of the values and visions of our place. Many are *home-blind* to the values and qualities in their place. Third, the depth of consequence relates to place-based planning, including an inquiry into what inhibits and enhances a sustainable *glocal* development. This includes a revitalisation of place-based culture and a widening of our sense of community, so to include the fellow beings we share the place with. Fourth, the depth of emersion gets at how we live our lives in the place, expressed through what we do in concrete situations. Again, these four depths are all interrelated, and we need to continue working on all of them.

A renewed sense of the sacred

The life philosophy I seek to express is inspired by being a Quaker (earthQuaker), and I find that there is in my life, my soul, a unique and personal pattern interconnected with the greater spirit and the greater body of God. In the book *Himmeljorden* (2009), I describe what I see as the unity between heaven and earth that we have direct access to through our spontaneous experiences. It awakens in us the mysterious unity in life; where there is no distinction between subject and object, map and territory, between me and the other. It is a spiritual or mystical experience of oneness with the world. We delve deep into the sacred and holy ground of being where our soul and the soul of God unite. It is closer to us than we think. The first Quaker, George Fox (1624–91), writes in 1648 after his revelation: 'Everything in creation gave another smell to me before, more so than words can express.' And the Quaker William Penn, who founded Pennsylvania, writes in 1693: 'How can humans find the conscience to misuse the Creation, while the Creator looks them directly in the face, in all and every part of it?'

I advocate a *panentheistic* view, where we answer to that presence of God in all living beings. I see an inner light, a creative, vital force – soul – in all that lives and moves. This inner light is related to what Spinoza, who had Quakers in his close circle of friends, calls *ratio*, the deeper voice of reason. This inner light would bring about a very different kind of Enlightenment of mind than that which was based on encyclopedic reason in Western culture.

The term *Heavenearth* is similar to what Abram, in *Becoming Animal* (2010), calls *Eairth*. Bringing the letter 'i' in the center of the word *earth*, the word *air* appears. The word *eairth* gets at how I am immersed in the air which is immersed in the earth. It is an expression that brings the spirit and body together in a sensuous whole. The air we breathe is the spirit that upholds our body like the atmosphere is the greater spirit that upholds the body of God.

I use the metaphor of *The Tree of Life* as an image of realising the unity between our soul and the soul of God. The tree has four interrelated parts: roots, trunk, branches and fruits, which I see as soul-awakening, soul-awareness, soul-consequence and soul-realisation. The roots of the tree are the mystical experiences of awakening to the unity of our soul with the soul of God. The trunk of the tree brings soul-awareness, where we become aware of the values and visions of the soul, our true Self. The branches show us the consequences the soul puts on our way of living in the various places we are at. The fruits of the tree are the realisations of the soul, which takes place when we live our lives in unity of spirit-body, heaven-earth, Self-God. Soul-realisation is when we live according to the inner light, *ratio*, the deeper reason, inner compass or the voice of God within.

A renewed sense of the sacred is born in the good, dark soil of being and the inner light helps it grow into existence as soul-realisation. We realise the unity of all beings, which will widen our understanding of the saying: 'Love thy neighbour as thy self', since the neighbours you share the place with, human and non-human, are ultimately part of your Self. A renewed sense of the sacred should furthermore be expressed in how we speak. If I speak of a tree as an object 'out there', I will relate to it very differently than if I speak to and with the tree as a unique, living being. It has to do with humility and awe to the unity that the tree and I share.

Sources of change

We have the knowledge and the technology to make fundamental changes in a green, just and peaceful direction, but still the changes are not taking place in our lives. I will here look at three sources that I see strong enough to help people make profound changes in their lives: fear, love and joy.

Fear is based on felt threat or crisis. Many will then change, but many will continue to live in denial and hope all goes well. It is as if our society is a large cruise ship that has just hit an iceberg and is taking in water. Some try to close the hole. Others yell to the captain to turn the direction the boat is heading in, if not it may hit more icebergs. Others are sitting in the restaurant, drinking wine while the water is reaching their knees, saying: 'Don't worry! The boat cannot sink!' Today many find ignorance to be bliss. Felt fear may get people to react quickly, but fear is a negative force we seek to avoid, and arguments intended to scare people, may lead to the opposite.

The other source is love. When two people fall in love, they are willing to do everything to care for one another. This is a positive force of attraction, which makes everything and everyone come into that light of love. But how can people fall in love and care for other living beings when research shows that people in the modern society use almost ninety per cent of their time indoors. It is difficult to love our neighbours, both humans and non-humans, when we never meet them.

Finally, there is deep joy. When I gathered texts by Naess on deep ecology over the last forty years, I was struck by the deep joy underlying it all. To live more in tune with our deeper values, following the inner light that enlightens our mind to our true Self, brings about a deep joy in us. We need to find our way into the changes where we are, what Naess called *sva marga* (*your way*, in Sanskrit), but at the same time, we are relational beings who are dependent on others to walk the way. Some would argue that profound changes will cause forsaking goods and going back in time, but a change of focus from standard of living to quality of life can be joyful and make us feel happier. Wealth is not in what you have, but in how you feel – you may have everything, but feel poor, or you may be poor, but feel rich. The deep joy of being immersed in this wonderful world, of living according to what one values, and of coming together with others, is the sustaining force that makes changes we seek to be wherever we are lasting.

References

Abram, David (2010) *Becoming Animal: An Earthly Cosmology,* Pantheon Books, NY.

Bowers, C.A. (1995) *Educating for an Ecologically Sustainable Culture,* SUNY, NY.

Børli, Hans (1992) *Med Øks og Lyre,* Aschehoug Books, Oslo.

Capra, Fritjof (1996) *The Web of Life,* Anchor Books, NY.

Carson, Rachel (1965) *The Sense of Wonder,* Harper and Row, NY.

Harding, Stephan (2006) *Animate Earth: Science, Intuition and Gaia*, Green Books, Devon.

Haukeland, P.I. (ed.) (2008) *Dyp Glede: Inn i Dypøkologien* (*Deep Joy: Into Deep Ecology)*, Flux Books, Oslo.

Haukeland, P.I. (2009) *Himmeljorden: Om det av Gud i naturen (Heavenearth: On the matter of God in Nature)*, Kvekerforlaget, Stavanger.

Naess, Arne with Haukeland, Per Ingvar (2003) *Life's Philosophy: Reason and Feeling in a Deeper World*, University of Georgia Press, Athens, Georgia.

Stoknes, Per Espen (2009) *Money and Soul*, Green Books, Devon.

3. Friendship for Life

SATISH KUMAR

Why is friendship important?

To my mind, we will not make much progress towards solving the global crisis unless each of us develops the basic attitude of friendship. Friendship for me is a supreme principle – the bread and butter of my life. I live by friendship. For me friendship is the most spiritual quality because friendship is unconditional, there are no ifs or buts, there is no reason why somebody is a friend. You don't say: I am your friend because you are this or that, you are educated, you are rich, you are intelligent, you are handsome, you are good to talk to, all those things don't come to your mind. You have friends because you are inspired to be a friend. And friendship is without any expectation. You don't expect anything, you just give and receive. Friendship has a deep gratitude but no expectation, only gratitude, only thanks. So in friendship you only say yes. When somebody asks you out of friendship you can't say no. So there is only yes in friendship. If somebody asks me some help out of friendship, I say yes. And if I ask someone out of friendship they say yes.

Friendship is not only for humans. I also feel friendship towards nature. I am a friend of my place and friend of my garden. I am a friend of trees, flowers. I am friend of the bees in my garden, and I am friend of even the earthworms and slugs and the snails in my garden. There is nothing in my garden which is not my friend. They are all my friends, even the weeds are my friends. The grass is my friend. So there is a kind of acceptance of everything as it is. You don't say: I like a garden when it is organised and designed in such and such a way. I like the garden as it is. So there is unconditional love and unconditional acceptance of my garden. So the garden is my friend. But my home is also my friend, because I accept the home as it is. Everything in my home is my friend.

Even my chairs are my friend. They look after me and I look after them. Even the glasses, the jars, the jugs, the knives, the forks, the spoons, the chopsticks and the cups from which I drink my tea, are my friends. So it's a very broad relationship.

And then it goes further and wider and I say: not only my home, my place and my garden is my friend, even my children are my friends. I don't say that my son Mukti and my daughter Maya are my son and daughter. In India we say that when your children become 16 they are no longer your children, they are your friends. It's a better term, friendship, than son and daughter, because son and daughter have expectation. You expect something from your children. They expect something from you as parents. As friends you don't expect anything. You treat them in a different way, in a respectful way. Same is with my wife June, she is my friend. That extends to everything. The village is my friend so I accept the village as it is. I don't say: my village should be like this or like that. I don't sit in judgment. There is no judgment. I accept the village, not only its people but the valleys and the trees and the natural landscape, I accept it as my friend. Then, I live by the ocean, and so the ocean is my friend. And I accept it as it is. Then it continues broadly, and I say the Earth is my friend, and the World is my friend. Whatever transformation I am trying to bring in my life, I'm trying to bring through friendship and love. Through acceptance.

So first of all I say the world is beautiful as it is, the world is wonderful, but within that world we have developed such institutions which need to be renewed. Like my home is my friend, but I clean it and from time to time I repair it, and paint it because after a while home becomes old and needs repair and renewal. And my garden needs repair and renewal. And in the same way, politics needs repair and renewal. So I bring transformation and renewal into politics. Economics needs renewal, so I work to have a renewal of economics. And farming systems need renewal, because they have been going for a long time and are in a state of disrepair. So I am bringing repair and renovation to systems of farming. It is all out of love, I don't have any judgment or criticism. Like my body needs washing and cleaning if it is dirty and healing if I become ill, if I become sick, if I have a headache, a skin problem or any problem in my body, I try to repair and renew and heal. In the same way, the world is my body, society is my body, and I know that society needs healing, repairing, renovating and renewing. So my work is the work of a healer, to bring renewal and repair.

That is how my work with the Small School is an act of friendship for children, my work for *Resurgence* is an act of friendship, to bring good values and good ideas to the readers. My work of Schumacher College is an act of friendship for the world. If I do something for my body, that is also an act of friendship for my body. Because my body needs healing and so I bring healing to my body, through meditation, good food, relaxation, rest, siesta ... When the body gets tired, I take a siesta, and that is healing. In the same way, I see society as my body. Sometimes society gets too frantic and works too hard, too fast, so I say to society: slow down. In the same way I say to myself and to my body: slow down, have a siesta. I say to society: have a siesta, slow down, don't work too fast or too hard. So it's out of friendship that I advise to my fellow human beings whether they are economists, or politicians, or business people, teachers or doctors, whoever they are I give them my friendly advice. If a friend asks me for advice, I give them advice about how to live more happily. I advise my friends to have less attachment, because attachment brings disappointment. Expectation brings disappointment. Live lightly and practise detachment, when you are detached you can keep moving, you're not stuck, there is no bondage. Detachment is to be free. All my work for ecology and for the environment and for spirituality, education, good agriculture is out of profound friendship in my heart for the world.

When I am working for the transformation of society and transformation of the world, I am working for the transformation of my own self. In this body I am a small self, but when I expand my consciousness then I identify with the greater self, which is the self of the universe. So the whole universe is my body. I am a microcosm of the macrocosm, of the universe.

How can we cultivate friendship?

Friendship must be embedded in the deep soil of relationship. Everything is related, we are all related to each other and to the earth. I was present in my genes and cells in the beginning of time, at the time of big bang. Ever since, everything that I am has changed forms in relationship to other genes and other cells and other life forms on the planet, and through the creativity and the consciousness we evolve. Life exists only in relationships. When seed is in relationship with the soil and the water, it grows, becomes a tree. When man and woman are in relationship, they produce a child. When I touch and sit in this chair, the chair holds me up. I am in relationship with the chair.

There is nothing I am not in relationship with. Everything exists in relationship, so my mantra is: we are all related to each other and to the earth. There is nothing outside relationship. The whole earth is our family. The earthworms and snails and bees, and wasps and mosquitoes and the birds in the sky and animals like deer and snakes, and lions and tigers and elephants, and trees and flowers and humans, the blacks and whites, the educated and non-educated, all are our family. We are all members of one family. We are all kith and kin, brothers and sisters, fathers and mothers. The American Indians always say: the Sky is my grandfather, and the Earth is my grandmother. St Francis of Assisi said the Sun is my brother and the Moon is my sister, and he talks to the wolf and says: wolf is related to me, and I have conversation with the wolf. He also says the ducks and its chicks are my family and I feed them. The humans and other than humans, they are all related. If there is no nature there are no humans. So we are related.

The Buddha puts the word *friendship* above all other words. When the Buddha was in his last days his disciples asked him in what form he would come back: would be he reborn as a great teacher or great prophet? And the Buddha told them he would not come back as a prophet or teacher but as *Maitreya.* Maitreya means friend. So, if we want to experience Buddhahood, we have to see all Nature as a friend. Whenever there is a spirit of friendship, the Buddha is present there. We meet the Buddha in friendship, in relationships.

But collective human arrogance deludes us into believing that we are the superior species; that we rule Nature, and so we can do what we like, because the Earth is ours. We can cut down the rainforests, pollute the rivers, over-fish the oceans, *conquer* Nature. The guiding philosophy of the last three hundred years (call it the Cartesian worldview, the Enlightenment Project, the notion of Pure Reason) tells us we can do what we like to Nature because it's all there *for us.* To value Nature only in terms of her usefulness to human beings is the symptom of a deep psychic disease. Now that Nature 'out there' is protesting, we have become fearful: fearful that we are all doomed. But if in reality we are all related and we are all friends then we ought not to be afraid of Nature. The Earth takes care of us, we should take care of the Earth, for we are one family; forests, rivers and animals are our kith and kin.

This arrogant mindset of human superiority over Nature has created an egocentric civilisation. This mindset is the fundamental cause of our current psychological, ecological and spiritual predicament. So we have to make a big shift – from this ego-centric worldview to an eco-centric worldview. *Eco* means relationship, and it comes from the Greek root *oikos,* which means home. Our earth is our home. And what is a home? A home is a place of relationship. When we go home we don't just go to bricks and mortar. It's relationships that turn a place into a home. *Oikos* implies a place of relationships. And so the eco-centred psyche is a psyche of relatedness and the ego-centred psyche is a psyche of separateness and isolation, of separation and superiority. We have put individualism on a very high pedestal: me, my benefit, my welfare, my ownership, my land, my house, my job; all this is egocentric. We need to shift from 'me' to 'we'.

I wrote a book in response to Descartes' idea of 'I think, therefore I am'. My book is called *You Are, Therefore I Am: A Declaration of Dependence*. As I speak to you now, who is it that makes me a speaker? It's not just that I have a few ideas to communicate, not just that I am able to put a few words of English together that makes me a speaker. If you were not here, I would not be here, I would not be a speaker! It's *you* who make me a speaker. The Earth is a reciprocal relationship. So the Earth is, therefore I am; my ancestors are, therefore I am; my parents ... my teachers ... like the Buddha and Mahatma Gandhi, Shakespeare, Bach and Beethoven are therefore I am. All those beings are in us. For we don't just drop out of a blue sky one day and become individuals, we carry the genes of our ancestors from millions and millions of years back.

'Who am I?' is a perennial question. Am I just my small identity: Satish; am I my secondary identities; like the owner of this house, the editor of this magazine, the writer of this book, with this skin colour, this religion, that political philosophy, that nationality. All these are in the sphere of ego-centred identity. But when I move beyond the sphere of small identities and I see that I am related to the universe and that the whole of the universe is in me, then I realise that I'm a part of the whole. Then I see that there's nothing out in the universe which is not in me. I am microcosm of macrocosm. The moment I expand my consciousness, the moment I am connected with the whole of the universe, I am able to touch the mind of God. Steven Hawking, in his book, *The Brief History of Time,* alludes to this idea in the last paragraph of his book. In my view knowing the mind of God is very simple. The moment you are *in* that eternal time you are touching the mind of God. By looking at history, objectively, intellectually or analytically you cannot touch the mind of God. You have to shift from geological time to dream time and see your place in it with your third eye. Then you touch the mind of God.

God is not somewhere out of this world, *separate* from this world. The God who created the world in six days and went to sleep on the seventh day is either still sleeping or, from somewhere he is pulling the strings that control everything. That is not my idea of God. For me creation is not a six-day event of the past; it is a continuous process and that is happening in every moment, and we are all co-creators and participants. This continuous creation is depicted metaphorically in Hindu iconography and terminology as Shiva dancing the universe. In the temples of Chennai and Chengalpattu dancing statues of Shiva in brass, bronze, copper, stone, wood, in many, many different materials, are created. That's the image of creation: the Universe as a dancing process. Evolution is a dance, going nowhere. Evolution is a dance not a linear process; it is a story not a linear history. A story, like dance, always goes around. Thomas Berry called it *The Story of the Universe,* not the history of the universe.

To connect with the Story of the Universe we need to be in Nature. But we, in our urban culture, are mostly stuck in the built environment, and we hardly have time to go out in Nature. The new generation of young people, are becoming more afraid of Nature. They say, 'I can't go out, it's raining. I can't go out, it's snowing. I can't go out, it's windy'. What's wrong with rain and snow and wind? Put a jacket on, it will be all right. Feeling, sensing and experiencing nature is a health imperative.

So when you are in the garden, you have a spade or a fork and you dig the soil, you plant a seed, a seed grows, you put a stake and you observe it and you look after it and you put water, compost, and with all that raw material of compost and water, soil, seed, fork, spade, when you bring them together in the right proportion, the right balance, the right harmony and attention and meditation, you turn that piece of land into a garden. That is a transformation.

Diversity and relationship

Good gardens are full of diversity. It's wonderful to have diversity. Evolution favours diversity. If there is no diversity, there is no evolution. If there is no diversity, how do we relate? So diversity is created for the purpose of relationship. And when we say we are related, then we can respect and have reverence for all life.

In relationship there is love. Whether you are living in Africa as a black person, or living in Europe as a white person, in China as a yellow person, as an American Indian as a red person, or you are speaking Swahili or Majorcan or Spanish or French or you follow Christianity or Buddhism or you are a communist or a capitalist – all these varieties are wonderful. Then we can love each other. Diversity is for relationship. I celebrate diversity because I celebrate relationship. Because we have forgotten the paramount importance of relationship, we dominate and exploit and kill and have hierarchy. We say something is higher and something is lower, and something can dominate something else. All this is the result of not accepting the world in its diversity, and not accepting the essential importance of relationship. Before we change or transform anything, we have to change our worldview and accept that we are all related. And we can bring transformation through relationship. Seed and soil in relationship can transform the seed into a tree. Teacher and child in relationship can transform a child a wise human being. We are all made of air, fire, water, earth, space, consciousness, creativity and imagination. These are the essential qualities from which everything is made. There is a unity of life manifesting in diversity. If you have a kilo of gold, you cannot use it so you transform it into fifty pieces of jewellery: a necklace, a bracelet, an anklet, a ring, you make different items of jewellery. They are shaped in different pieces, but they are made with the same gold. In the same way, we are all made of the same basic elements. If you have just one element, one kilo of gold, then there can be no relationship. Only by transforming one into many can you create relationship.

Is there something deeper than relationship?

Love is the foundation upon which all other things I have talked about are built. Friendship is built on love, relationship is based in love. And love is the greatest power. When you have love, then fear evaporates and eventually disappears. We need to live and act in our life from the basis of love.

I act in my life from the basis of love. I am an environmentalist not because I have fear of doom and gloom, or disaster or end of civilisation. No, I am not driven by fear. My social movement, environmental movement, spiritual movement, of which I am a part, is driven by love. My ecology is driven by love. I love the trees, I love the forest, I love the mountain, the soil, the colours of the flowers, the smell of the fruit, the sound of the wind. I love the sunshine, the rain, I love the snow, and I love human beings, I love communities. I love children, I love women.

I think when there is attraction in nature, the plants are attracted towards the sunlight, there is a love between the plants and the sunlight. Love is not only a human attribute, love is an essential ingredient in nature. When bees pollinate the plants, that is an act of love. When there is a mating in all creatures, to reproduce and procreate, that is love. Physical love is also part of the spiritual love and attraction, it's a kind of allurement, it is the biology of love.

There is also the economics of love. When we make something we want to serve the community. Essentially, the farmer is growing food out of love for the community and the family. There is love there. When a shoe maker is making shoes, there is love in every stage, love of making shoes, and then love for the person who is going to wear the shoes. When it is made with love, it's a beautiful shoe. Because it's made with love. And when those pair of shoes made with love, they are more comfortable shoes. When you make something with love, you make it very lovely, very perfect, very beautiful. You are not cutting corners, you are not economising. You put every effort, your best ability to make the best shoe because of love. That example can go with everything. When you make something with love, then it is very beautiful. When you make something with love, the making itself has intrinsic value and it is an end in itself. You are not doing something to achieve something else, not working towards an outcome. You are not attached to the fruit of your action, because your action itself is an act of love. When I am talking to you now, this itself is an act of love. I have no desires to achieve something, to have an outcome or to get something. There is no exterior motive, the only motive is the act of love itself. This is pure action. Every act is an act of love. There are no other attachments. So for me, love is a process of living. Love is means and the end. Love is the way and also the place of arrival. There is no distinction. There is no way to love, love is the way. I am falling in love every moment. The moment I wake up, I am falling in love with the beautiful sky and sunrise, and I am falling in love with my beautiful wife every day. And I am falling in love with my parents, and children, every day fresh love. I live in the house of love, in the city of love. Life becomes very light when it is led with love. There is no burden, there is no bondage. When you act out of love, there is no karma. If you live out of love, you accumulate no karma. Because your action is so pure, it does not stick, like a feather, so light. Gone. And when you are in love, you are not expecting anything. You have no desires, it's love for the sake of loving. There is no other destination, love is the destination. There are many levels, personal love, social love, universal love. Love is such an all encompassing word. I love potatoes, I love the universe, I love God. There is no other word so full of meaning. Out of love, you discover a sense of the sacred. Out of love, you sacrifice your comfort to maintain life, to maintain family, to maintain your relationship. The word sacred comes out of sacrifice. Sacrifice is not borne out of hardship, or reluctance, that is not sacrifice, it's not love. True sacrifice is no burden, you don't suffer, you don't even notice that you make a sacrifice. In sacred love, you go all the way, there is no limit. It's unlimited and unconditional. You don't say I come this far, if you come this far. That is not unconditional love. Sacred love is without limit. You are prepared to give your life. You are prepared to die for love. And you don't even know that you are dying for love, you just die.

References

Hawking, S. (1995) *A Brief History of Time: From the Big Bang to Black Holes,* Bantam Books, UK.

Kumar, S. (2000) *No Destination,* (second ed.) Green Books, Dartington, UK.

—, (2002) *You Are, Therefore I Am – A Declaration of Dependence,* Green Books, Dartington, UK.

4. Gaia: From Story of Origin to Universe Story

JULES CASHFORD

With the splitting of the atom, everything has changed save our
mode of thinking, and thus we drift towards
unparalleled catastrophes.

ALBERT EINSTEIN

The old gods are dead or dying and people everywhere are
searching, asking: What is the new mythology to be, the mythology
of this unified earth as of one harmonious being?

JOSEPH CAMPBELL

Earth, is it not just this that you want: to arise
Invisibly in us? Is not your dream
To be one day invisible? Earth! Invisible!
What is your urgent command, if not transformation?

RAINER MARIA RILKE

We are all born into a story – the story of our family – and, as we grow, our story grows with us. It opens out into our local community, then later into our tribe, our race, our country, our species, and the age in which we live: the story of our time. But no story is complete without the ultimate story of the Universe, which is, of course, the primary story: the story of origin of *every* family of the Universe – non-human as well as human – from which all the other stories take their reference and meaning.

Stories of Origin, or Myths of Creation, as they are also called, belong to every culture in every age. These are sacred stories which explore a vision of the whole Universe and the place of human beings, and all other beings, within it. They are stories of wonder and celebration and gratitude, fostering harmony between people and the Universe, between the microcosm and the macrocosm, the part and the whole. They ask for understanding and guidance as to how to be in the world that comes to us as a gift of life.

All these stories are fundamentally a search for dialogue, a quest for relationship with the mystery of the whole which surpasses and encloses us. As mythic images they have a universal dimension common to all human beings by virtue of being human, which is why they are recognisable to all of us; and they have also a local, ethnic dimension, particular and specific to each person, tribe, race and place. The different kinds of answers to these questions all over the world are then central as to how the people within their own unique culture are going to live and what they will value.

What happens, then, when a Story of the Universe becomes fixed in the past, and cannot grow and change when the conditions of life and the needs of the time have grown and changed? What happens when the old story can only answer the old questions, first asked long ago – in a different world – even as long as two thousand years ago? And when new questions cannot be heard because the old story rules them out? Inevitably, the old Universe Story loses its magical wonder and its sense of the infinite, and so can no longer guide and inspire.

If this is so, might we not anticipate a fundamental dissonance afflicting people's lives and values? For when the original harmony in feeling between the part and the whole is lost, might this dissonance not manifest itself as alienation and helplessness, and, worse, denial of the destructive effects of the old ways of thinking, which then continue as before, even though we can see the consequences? There must always be an interim stage in the process of change when we try to break free from the constraints of the old but have not yet been completely captured by the radical call of the new. This is the time when the visionaries among us call for a new way of being, but others – often too invested in the rewards of the old way of thinking – cannot or will not hear. For the new story may demand sacrifice from us, not least the sacrifice of the old story to which we had become accustomed. Then we are left *between stories*, cut off from the deepest roots of our being, which relate us to all life.

In our time, mostly in the west and the north of the planet, we can see the devastation of the natural world all around us. It must therefore be that, with the exception of Indigenous People, Nature is rarely experienced as sacred. When we look for the revelation of the divine in Nature, in cultural values enshrined in custom and law, we do not find it.

No *absolute inviolable laws,* which are everywhere the sign of the sacred, exist to protect Nature from the excesses of human behaviour. That can only mean that the present Universe Story has become partial and can no longer summon the best from all human beings, *precisely because* it disregards their living context of Planet Earth. It must assume that the beauty, inspiration, teaching and healing of Nature are so inferior to its own as to be disregarded, frequently under the guise of 'progress', which is to say, advancement for humans on human terms, no matter what the suffering to other life forms with whom we share the planet. We can only conclude that the present Universe Story has lost its universality; it stands only for the human, not for all the life forms of Nature, many of whom are becoming almost daily extinct. It can no longer comprehend the whole of life both actual and potential, within and beyond our understanding.

The dominant and most prevalent Story of the Universe in the west is the Judaeo-Christian story which has structured and informed our ways of thinking for the last two thousand years. We may think of Christianity as a religion which inspires many lives with devotion and value, primarily relevant to those who believe in it. But the psychologist C.G. Jung suggests that this is to underestimate its fundamental influence in western habits of thought and action, even when we are not fully aware of it. He introduces the important idea of 'the inertia of the unconscious' to point to the ways in which we may adopt a new position intellectually, but overlook the deeper levels in the psyche which resist change:

> We think we have only to declare an accepted article of faith incorrect and invalid, and we shall be psychologically rid of all the traditional effects of Christianity and Judaism. We believe in enlightenmnet, as if an intellectual change of front somehow had a profounder influence on the emotional processes or even on the unconscious. We entirely forget that the religion of the past two thousand years is a psychological attitude, a definite form and manner of adaptation to the world without and within, that lays down a definite cultural pattern and creates an atmosphere which remains wholly uninfluenced by intellectual denials. The change of front is, of course, symptomatic, but on the deeper levels the psyche continues to work for a long time in the old attitude, in accordance with the laws of psychic inertia.[1]

When we study the Judaeo-Christian story more closely – at least as it has been *literally* interpreted in the west – we can see that its focus has been upon human beings alone as potentially (but only inadequately) reflecting the image of divinity. By contrast, the Earth, and all beings on Earth, have been, in strict doctrine, excluded from divinity. In Genesis, the first book of the Bible which contains the Judaic stories of creation, humanity is drawn as fallen because of disobedience to God (interpreted in later Christian theology as 'Original Sin'), and the Earth is cursed as a consequence – 'for thy sake'. (In fact this is only one of three creation stories, but the one that came to stand for the others.) Those who read this literally, as many people over the millennia have done and still do, see a god *transcendent* to creation – to Earth and Nature and Human Nature, and indeed the Cosmos as a whole. This means that, taken as literal doctrine, divinity is located *beyond* creation; it is not immanent ('remaining in, or inherent') *in* or *as* creation.[2] When Spinoza, a seventeenth century Dutch philosopher of Portuguese Jewish origin, said that: 'God and Nature are two terms for the same substance', he was talking of the divine as *immanent* in Nature. Just as the poet William Blake proclaims that: 'Everything that lives is holy'.

It is hard to ignore the correspondence between the *absence* of the Sacred Earth in the Judaeo-Christian tradition and the *desacralisation* of Nature in the daily actions of many human beings who come from this tradition. Certainly, there is no register of the *inherent* sacredness of Earth in Judaeo-Christian law. Now that the focus on the human project, inherent in the old story, has reached its peak and demonstrated its terrifying limits, as Einstein shows, it is possible that a new story is the only thing that will save the planet *from* (not, as the human-centred vision of the world might have it, *for*) humanity.

But how does a new Universe Story come into the hearts of human beings? Would we even recognise it when it arrived? And what if it were already here, waiting to be acknowledged?

The image and idea of 'Gaia' appears to have entered some portions of western consciousness in the last forty years or so, and since then to have acquired a life of its own. This is just what we might expect from an autonomous image which appears to arise spontaneously from a region beyond our rational control. This level, or layer, or depth, of the psyche has been variously named: Plato called it the Soul of the World, *Psyche tou Cosmou*; the Neoplatonists called it the *Anima Mundi*; William Blake, the first of the Romantic poets, called it the Poetic Genius; Samuel Coleridge, also a Romantic poet, called it the Primary Imagination; the Irish poet, W.B. Yeats, called it the Great Memory and the Great Mind; Jung called it the Collective Unconscious; Many people talk of the Creative Psyche, the Otherness which transcends the personal psyche; and, as in earlier times, it may be simply felt to be the sphere of revelation of 'the divine', expressed all over the world through the gods and goddesses of many names. It may be called 'the Ancestors'. For the South African people of Venda, for instance, it is *Mupo*.

The apparent coincidence which brought the image of Gaia back from the distant past was a random walk taken by the scientist James Lovelock in the late 1970s with his friend William Golding, who was a neighbour and the author of a popular novel, *Lord of the Flies,* and also a classicist. Lovelock was looking for a name for his new hypothesis that the Earth was a self-regulating system. He wanted to propose that the Earth had the capacity for homeostasis, that is, for comprehensive inner adjustment and self-regulation in response to changes in the outer world. Golding suggested 'Gaia', the name of the ancient Greek Mother Goddess of All, and the 'Gaia Hypothesis' was born.[3]

But no-one had anticipated that the image of Gaia would catch the Imagination of the time, almost as though it had a mind of its own. The 'Gaia Hypothesis' soon became more radical than Lovelock had intended, as though the hypothesis which excited less scientific minds than his own were primarily to do with 'Gaia herself', with a vision of Earth as alive – not, as in the old story, dead 'matter', to be governed and explained only by mechanistic laws. This 'new Gaia' called forth the 'old Gaia': they were both animate, intelligent, purposeful – in a word, *ensouled* – more like a Goddess than a Machine.

In retrospect, it seems as if what appeared as coincidence – of the name of 'Gaia' to describe a scientific hypothesis otherwise destined to be known as 'systems theory' – could be meaningful at a deeper level. Jung proposed the idea of 'Synchronicity' to draw attention to the way that certain events, happening together at the same time, may be connected not by cause but by meaning. In which case, the intuition of meaning may refer us to a deeper realm which underlies both physical and psychological expressions. 'In the depths,' the poet Rilke says, 'all becomes law'.

We might expect a new Universe Story to be manifest in a new kind of numinous image, one which beckons us beyond our habitual categories of perception – '*translucent*', in Coleridge's term, to 'the eternal in and through the temporal'. It might be numinous precisely because it had come from the depths to compensate for the present imbalance in human consciousness, trying, in its image of a living Earth, to restore harmony – that harmony which was Plato's image both of the universe and of the individual soul in perfect 'attunement' with it (*harmozein*). A numinous image may also be an old image seen in a new way – such as 'Gaia' – one which leads beyond the merely human into the totality of the universe, and invites us to celebrate an organic living whole where before we have focused on our difference from – and superiority to – all other beings.

Could it be that the old image of Gaia has begun to speak to us in a new voice, as though in answer to the crisis and opportunity of our time?

In our present crisis of 'environment' – a barren term to refer to Nature which means, from the French *environs*, merely 'round about' – the name of Gaia is frequently heard. 'Gaia' has come for many to embody a new kind of consciousness, sometimes even called 'Gaia Consciousness', which expresses a reverence for the planet as a living whole who is home to all other living beings, *all* of whom share in and give form to her own original and dynamically changing life. In this vision, Earth becomes again, as she used to be, a Living Presence, a Subject, and a 'Thou' who requires from us relationship – in a word, sacred. In this vision, Earth can no longer be seen as dead matter, an object, an 'It' – merely a resource for humans to plunder as they will. But where did 'Gaia' come from?

Gaia in ancient Greece

Gaia was one of four stories of origin in ancient Greece, otherwise known as a creation myth. To understand the crucial importance of myth for the ancient mind, and also for modern consciousness, we need to appreciate the awe and respect which ancient Greeks accorded to their myths. The word 'myth', from the Greek *muthos*, means story or speech, but one which sets a pattern and has purpose and design within it, deriving ultimately from the Indo-European root of *mud*, meaning 'to think' and 'to imagine'. In early Greek thought *muthos* came first and was later contrasted with *logos* which arose out of it and meant word, speech, statement, account, thought, and reason – from which all our 'ologies' come: *mythology,* the logos of myth; *psychology,* the logos of the psyche; *anthropology,* the logos of humans. Logos became *ratio* in Latin, which was usually translated as 'reason' alone, and so lost its original complexity. Briefly, then, a myth is a story inspired by Imagination, while *logos* is an account answerable to Reason.

In ancient Greece, it was held to be crucially important to have a balance between these two ways of understanding the world, accepting that both were necessary. They embodied two different but complementary ways of knowing the world, and so vital was this distinction that there were two different words for 'knowledge'. *Gnosis* was knowledge won through relationship and love: the way you know a person, an animal, a plant, a garden, landscape, poem, story or myth. It engages the whole personality, in contrast to *episteme,* which was knowledge *about* something, a term which gives us 'epistemology'. It is a way of knowing which engages primarily the rational mind.

We might think of this contrast as the difference between poetry and prose, the poetic image and the reasoned statement. However, with the rise of science in seventeenth century Europe and the increasing influence of the industrial mind, Logos has become supreme, as has the rational way of knowing which goes with it, and the balance between Logos and Mythos has been lost. Indeed, 'epistemic knowledge' has been increasingly valued *just because* it leaves out the 'subjective' experience, calling itself 'objective'. As a result, 'myth' is often dismissed as a lie, fantasy or illusion, just as 'story' becomes a tale for children, something arbitrary and not necessarily 'true'. Consequently, it is imperative today to reclaim the value of story and image, and the imaginative way of knowing that they invoke, as against the claim of Reason to dominate and be sufficient on its own, dismissing the Imagination as not equally essential to human thinking.

On the contrary, some would say Imagination is the ultimate in human thinking, and has laws of its own no less rigorous than those of logic. Reason, says Blake, is only 'the ratio of what is already known'; unlike instinct, passion and feeling, it can create nothing from itself; whereas, 'To a Man of Imagination, Nature is Imagination itself. As a man is, so he sees'.[4] Jung memorably wrote that: 'hemmed round by rationalistic walls, we are cut off from the eternity of Nature'. Einstein often said: 'The best thing is Intuition'. While Thomas Berry, the Passionist monk who was a cultural historian and ecological theologian, or, as he preferred to describe himself, 'geologian', urgently proclaimed: 'Loss of Imagination and loss of Nature, they are the same thing'. Elsewhere, he continues:

> If the beauty of the land is disfigured, if the fertility of the soil is lost, if the rivers are polluted, if living species begin to disappear, then the integral life of humans is endangered. The human soul begins to shrivel. The Imagination becomes either dessicated or distorted.[5]

If we follow the image and story of Gaia in ancient Greece, it may be easier to see why now, in the present, we have to look back into the past for an image to carry us forward into the future. For Gaia was *both* the name of the Great Mother Goddess Earth *and* the everyday Greek word for earth – only the capital letter distinguished them. This suggests that the physical form of Earth – the ground we stand upon, the soil we plough – took its reality from the sacred realm, which we acknowledge by understanding Earth as our Mother and Mother of all. The two terms were permeable to each other, so the early Greek mind could move fluidly between them without having to reach for a different kind of language to explain 'which one' was meant: they were one and the same. The 'Earth that gives us grain' was also the 'Mother who feeds the world', who was 'the oldest one', the foundation, the 'Mother of the gods'. So Gaia, as Goddess, ground and globe, was always transparent to the deeper poetic vision – *Zoon* was Plato's term for Earth – a Living Being.

In ancient Greece, there were four different Stories of Origin, each reflecting the different groups of people who had come to Greece at different times and from different places: the native Pelasgian, the Orphic, the Homeric, and the Olympian. In different ways they all imagine the universe in the image of a Great Mother Goddess who brings forth creation from herself. Gaia belongs to the Olympian myth of creation. This *Homeric Hymn to Gaia* was written down in 500 BC, though probably chanted or sung at festivals for many centuries before that:

> Gaia, mother of all,
> the oldest one, the foundation,
> I shall sing to Earth.
>
> She feeds everyone in the world.

Whoever you are,
whether you walk upon her sacred ground
or move through the paths of the sea,
you who fly,
it is she who nourishes you
from her treasure-store.

Queen of Earth, through you
beautiful children,
beautiful harvests,
come.

You give life
and you take life away.
Blessed are those you honour with a willing heart.
They who have this have everything.

Their fields thicken with bright corn,
the cattle grow heavy in the pastures,
their house brims over with good things.

The men are masters of their city,
the laws are just,
the women are fair,
happiness and fortune richly follow them.

Their sons delight in the ecstasy of youth.
Their daughters play,
they dance in the grass,
skipping in and out,
they dance in the grass over soft flowers

It was you who honoured them,
generous goddess, sacred spirit.

Farewell, mother of the gods,
bride of starry Heaven.

For my song, allow me a life
my heart loves.

And now and in another song
I shall remember you.[6]

The poet Hesiod, writing two hundred years earlier, was the first to write a long poem dramatising the whole process of creation, called *Theogony*, the 'Genealogy of the Gods'. After an invocation to the Muses he begins:

> Chaos was first of all, but next appeared
> Broad-breasted Gaia, sure standing place for all
> The gods who live on snowy Olympus' peak,
> And misty Tartarus, in a recess
> Of broad-pathed earth, and Eros, most beautiful
> Of all the deathless gods ...[7]

Tartarus (Underworld) and Eros (Love) appear after Gaia has emerged from Chaos, as though her emergence releases the structural principles of the universe and the web of relationships which binds them together (Eros). Chaos (sometimes called Chasm since the Greek does not contain our later ideas of confusion or disorder), then produces Night and Erebos (Darkness), and Night in turn, uniting with Erebos, gives birth to Day and Space. As though independent of these more abstract structuring principles, Gaia gives birth out of herself (the original meaning of 'Virgin') to Heaven (Ouranos), hills and mountains, and Sea (Pontus):

> And Gaia bore starry Heaven, first, to be
> An equal to herself, to cover her
> All over, and to be a resting-place,
> Always secure, for all the blessed gods.
> Then she brought forth long hills, the lovely homes
> Of goddesses, the Nymphs who live among
> The mountain clefts. Then without pleasant love
> She bore the barren sea with its swollen waves,
> Pontus. And then she lay with Heaven ...

Transforming her son into her lover, in the widespread tradition of son-lovers of the Goddess, Gaia unites with Ouranos, Heaven, and together they give birth to the next generation of divinities called the Titans: six goddesses and six gods, among whom were Mnemosyne (Memory), Themis (Law), Rhea (the Flowing One), Phoebe (Moon), Hyperion (Sun) and Chronos (Time).

We can see that Gaia was an image or an idea far beyond what we generally mean by 'Earth': she was the Mother who brought forth the universe from herself, so all children were children of the universe, formed from her substance. Earth, then, carried the memory of the whole (Mnemosyne), was inherently lawful (Themis), embodying the changing rhythms of time. It followed that lawfulness and memory and temporal rhythm also belonged to all her creation. In contemporary terms, Gaia was a vision of the universe as one dynamic living whole.

As we follow the story of Gaia unfolding into creation it is remarkable how Gaia herself was understood as the dynamic force of each stage of creation. Inevitably, the stages of the Earth's procreation, as imagined by humans – Gaia giving birth to children who give birth to their children and so on – are also the stages of differentiation of human consciousness. However, in the Imagination of ancient Greece, Gaia was never left behind as a static legend of the beginning of things. Gaia continued to play an essential role whenever the drive of creation was required to reach a new level – inaugurating, for instance, the cyclical movement of the Seasons – offering an original model of the self-regulating process of which Lovelock speaks. Aeschylus, the first great Greek dramatist, writes:

> Yea, summon Gaia, who brings all things to life
> And rears and takes again into her womb.[8]

It followed that Gaia's law could be profoundly disturbed by the unlawful and immoral behaviour of human beings. Sophocles, in his *Oedipus Rex*, draws an Earth intimately related to the moral life of humanity. Oedipus, the King, is quite content in his unconsciousness until Earth suffers. It is Gaia's protest which initiates the drama of Oedipus' awakening to what he has done and who he is: the slaying of his father and the marrying of his mother. Suddenly, the land of Thebes begins to die:

> A blight is on the fruitful plants of the earth,
> A blight is on the cattle in the fields,
> A blight is on our women that no children
> are born to them.[9]

Oedipus sends to the Delphic Oracle of Phoebus Apollo to reveal the cause. And the oracle, whose first law was 'Know thyself' and whose second law was 'Nothing in excess', now defines for all time the meaning of pollution as a human crime against the divine order, the profaning of what is sacred:

> King Phoebus in plain words commanded us
> to drive out a pollution from our land,
> pollution grown ingrained within the land.

Here, the pollution is the presence of the murderer of Laius, the former king and Oedipus' father. When Oedipus discovers that he is himself the pollution and leaves the city, harmony between the human and divine order is restored and Earth comes back to life. The drama shows that the laws of Gaia are written into all creation.

With so rich and complex a tradition who would have suspected that Gaia was to be the last Goddess of Earth in the West, the last time that the Earth was formally revered as sacred? Even to talk, then, of the sacredness of Earth is to press against the weight of two thousand years of religious history, which permeate cultural and psychological assumptions about the nature of reality. To reach for the image of Gaia to heal this division in the western psyche might seem a folorn hope, if it were not that 'Gaia' is, and has always been, a universal idea.

Gaia in ancient India

For Gaia does not belong solely to the classical western tradition, which in any case is suffused with the language and thought of the earlier Indo-European cultures. The actual name 'Gaia' comes from India at least one thousand years before – *Gaya* in Sanskrit – where it appears in the *Vedas* and *Upanishads* in the form of the *Gayatri Mantra* which has a meaning that also expands infinitely to include Earth, humanity and all other beings. In a sense it is also a Story of Origin or a Creation Myth which relates humanity to Earth as the image or song of the whole. If there were one phrase to capture the meaning of Gaya it would be 'Moving Song'.[10]

This understanding is reflected in the Goddess of a Thousand Names, of whom Apuleius speaks in the first century AD,[11] found all over the world in Bronze Age, Neolithic and even Paleolithic times, persisting in the west to the Romans and beyond, and in the east and south never completely dying out – the image which expresses the feeling of early people that Earth is the Universal Mother, such that the human story and the universe story are one and the same.

Gaia as a symbol of a new universe story

For many of our twenty-first century minds, Gaia is not now so much a goddess as a symbol, but we can still ask: a symbol of what? Even in olden days, goddesses and gods were not worshipped exclusively as ruling over different areas of natural and human activity, like queens and kings; they were often more seriously understood as ways in which the world revealed itself.

Similarly, we could infer that the world has revealed a new dimension of itself in 'Gaia'. Symbols announce themselves; they are not 'made' by us; all we can do is recognise them, dream the dream onwards in new dress, as Jung puts it. Is it possible that Gaia has revealed to us that we are in the process of that transformation of which Rilke speaks? 'Earth! Invisible! What is your urgent command, if not transformation?' Could Gaia be becoming a symbol for a transformed vision which implies a new relationship to the whole of life and a different way of understanding our role in the Earth's own process of transformation?

However, the entrance of the image of Gaia in our time is not so sudden and unexpected as it might at first seem. It rests on a sure foundation that has apparently been long preparing itself for change, certainly since the 1960s with the emerging of Feminism and the Green movement, and probably, though rather too slowly, for the last two thousand years – ever since, in fact, the divinity of Earth was banished, first by the Babylonians in c. 2000 BC (in the recreation of the world by the god Marduk from the dead body of the original Mother Goddess Tiamat), and then by Judaeo-Christianity, when the feminine principle was doctrinally excluded from the nature of the divine.

For, to reiterate, in the Judaeo-Christian creation myth, with its transcendent god, Yahweh, not immanent in his creation, Earth was cursed for the 'sins' of humanity: 'Cursed is the ground for thy sake ... in the sweat of thy face shalt thou eat bread, till thou return unto the ground; for out of it wast thou taken: for dust thou art, and unto dust shalt thou return' (Genesis 3:17–19). Strictly, for some elements of the early Christian Church, including St Augustine, both Nature and Human Nature were fallen.

The problem seems to be perennially one of symbolism and metaphor, which belong to the Imagination and point us to *inner* realities, as opposed to literal and historical fact, which are claimed by reason and subjected to *external* tests of empirical verification – no less partial, as we have seen in the long history of religious strife, than any other activity. For, as Campbell says, whenever the images of myth are taken literally or historically they are killed, and so prevent or obscure their traditional role, which has always been to carry the human psyche forward across those difficult thresholds of transformation which require a change in the patterns of unconscious as well as unconscious life.[12]

So when a transcendent god, from whatever culture, is understood as literally and necessarily outside or beyond creation, a polarisation comes into the way of thinking about the Earth which deprives Earth, formerly Mother Goddess Earth, of her original divinity. This takes the form of a new polarity of Spirit and Nature – Spirit now belonging to the god transcendent to creation, and Nature belonging to the creation which is now deprived of Spirit. Yet 'Spirit' and 'Nature' were originally inherent in Earth as one phenomenon – the two terms were as yet indistinguishable because both were always present. More precisely, there was no need for two terms to be first distinguished and then polarised, because they were essentially one and the same. Gaia was Animate Ensouled Earth.[13]

But when, as in the collective cultural tradition of the west, the awe instinctively due to the sacred mystery of Earth is transferred to a transcendent god and to 'the race of man' in his image and in his name, then Earth, suddenly barren of love, becomes an object of fear, a chaos to be 'dominated' and 'subdued', by human beings, or an inert mass to be breathed into from above. Such was the effect of Yahweh's curse upon Earth: nature and spirit split apart – nature was fallen, earth was dust and spirit was transcendent, belonging first to the god (without the goddess) and then to the man (without the woman). Oppositions multiply, and so – in the literal language of dogma – instincts, feeling, intuition and Imagination were assigned to 'the feminine' and the female (who was 'more natural', which is to say, inferior) and conceptualisation, thinking and reason were assigned to 'the masculine' and the male (who was 'more spiritual', that is, superior).

If this seems extreme, consider Eve, who was held to be a secondary creation and so of inferior substance and so the 'devil's gateway' for sin and death. 'But I suffer not a woman to teach, nor to usurp authority over the man, but to be in silence. For Adam was first formed, then Eve.

And Adam was not deceived, but the woman being deceived was in the transgression' (St Paul, 1Tim. 2:8–14). Or, as Martin Luther, champion of the Protestant Reformation, put it, 'If the serpent had approached Adam it would have been a different story'.

However, what is less well known is the response of the Collective Unconscious or the Soul of the World to rectify this imbalance, working apparently in the same way as the individual unconscious works to restore harmony to a conscious point of view that has become one-sided and so diminishing and ultimately destructive. We can trace it, for example, in the call of many thousands of people, millions even, for the Virgin Mary to play a more essential role in the imagining of the divine Christian order. From her modest appearance in the New Testament, first as the pure vessel and second as a simple loving mother, she has become astonishingly close to being elevated into divinity, to becoming, in image if not doctrine, 'a goddess'.

For, in Christian thought, Mary, Mother of Jesus, has symbolically inherited most of the imaginative range of the earlier Mother Goddesses – she is 'Queen of Heaven', 'Star of the Sea', 'Queen of the Underworld' ('Now and at the hour of our death'). But she is not – nor, doctrinally, could she be – 'Queen of Earth'.

In various Councils and Papal Bulls from the fifth century AD onwards, Mary has been called 'God-bearer', *Theotokos* (AD 431), 'Ever-Virgin', *Aeiparthenos* (AD 451), unable to die (AD 600) – the Feast of the Dormition, or Going to Sleep, was declared a public holiday on August 15, the day that many centuries later was to celebrate her Assumption into Heaven. From 1170–1270 in France alone one hundred churches and eighty cathedrals were built in her honour. In 1854, she was declared Immaculately Conceived, as well as Immaculately Conceiving. In 1950 her Assumption into Heaven was declared, and in 1954 she was proclaimed Queen of Heaven to the thunderous applause of eight million people in the square in front of St Peter's Basilica. But because the theology in which the image of Mary was enclosed had not changed, divinity was again measured by its distance *from* Earth, not its epiphany *as* Earth, so the old antinomies were simply reasserted at a higher level – and the need for an image outside the Christian tradition remained as great as before.

Yet why should it matter what status Mary has? The fact that it did matter to so many people all over the world appears to validate Jung's exciting claim that: 'In the Collective Unconscious of the individual, history prepares itself'.[14] If we consider that we also are 'Gaia' – some would say the Earth conscious of itself, others would say just one of the infinite forms of the consciousness of Earth – then is it not possible that, in ways we can barely conceive, this could also be the self-regulation of the whole, suggested in the 'Gaia Hypothesis'? That, in other words, 'Gaia' works through the Collective Unconscious, or perhaps, more imaginatively, Gaia is the latest expression of the dynamic Soul of the World?

In the long journey of humankind from its earliest beginnings, we might expect that images of a unified world become real at a different level of understanding, so that what was once belief becomes metaphor – ultimately a symbol of a deeper reality than we can reach with our minds alone. The image of the universe as an unbroken wholeness, composed of a web of relationships, containing an ocean of energy, having an implicate as well as explicate order, being in a continual process of movement with no absolute point of rest and changed by our relationship to it – these are images from modern sub-atomic physics. Whereas in the myth of the Goddess, these images were believed to be true because all life was of the substance of the Goddess, she who was everywhere and everything, binding all things together in a great web of life, woven by the spinning goddesses of the Moon, continually manifesting in a thousand images – ocean, sea, heavens and earth, and the creatures therein – both visible and invisible. However, the language of the New Science might remind us that all the great mystic teachers have had a holistic vision, embodied in a passion for right living, which extends to all living beings: the notion of Buddha consciousness in all things, the Hindu vision of *Tat Tuam Asi*, 'Thou art That', and the words of Jesus in the Gnostic *Gospel According to Thomas*, found in an urn in Nag Hammadi in 1946, and hidden for two thousand years:

> Cleave (a piece of) wood, I am there;
> lift up the stone and you will find Me there.[15]

This vision evokes the words of the greatest of our contemporary scientists, Albert Einstein:

> A human being is part of the whole called by us 'the universe', a part limited in time and space. We experience

> ourselves, our thoughts and feelings, as something separate from the rest – a kind of optical illusion of our consciousness. This delusion is a kind of prison for us, restricting us to our personal desires and affection for a few persons nearest to us. Our task must be to free ourselves from this prison by widening our circle of understanding and compassion to embrace all living creatures the whole of Nature in its beauty.[16]

To conclude, in our time, the apocalyptic image of Earth seen from the Moon may have implicitly invited the unconscious of human beings to acknowledge that Earth has character, intelligence, soul: in a word, a consciousness of its own. This image of the Earth as a whole entered the collective Imagination in 1963, and its repercussions have probably only just begun.[17] We all know that from the moment that the Earth could be seen from the Moon – looking like the Moon has always looked from the Earth – a new relation to Earth was inevitable. Never before had we been able to see with our eyes what we already knew with our minds – that Earth was set in space and spun round the Sun; and never before had we been given a perspective beyond Earth from which to view Earth so that we could for the first time in human history see our planet as a whole, 'entire in itself', as a totality, but also finite, limited – even, against the vast black backdrop of space, intensely fragile. Yet, this numinous vision of 'a new heaven and a new earth' cannot but move us into a new mode of being: the consciousness of humanity participating in the consciousness of Earth.

And the name for this Story of the Universe, both very new and very old, as old as Earth, would seem to have announced itself as – 'Gaia'.

Notes and references

1. C.G. Jung, *Collected Works* (1957-79) eds., Herbert Read, Gerhard Adler, Michael Fordham, William McGuire, trans. R.F.C. Hull, London, Routledge & Kegan Paul, Vol. 6, p.185. See also Jung, *CW,* Vol. 8, p.381.

2. For further discussion, see Baring, Anne and Cashford, Jules (1992) *The Myth of the Goddess: Evolution of an Image*, Penguin, London.

3. James Lovelock (1979) *Gaia: A New Look at Life on Earth*, OUP, Oxford.

4. William Blake (1961) Letter to the Rev Dr Trusler, 1799, in *Complete Poetry and Prose of William Blake*, ed. Geoffrey Keynes, Nonesuch Press, London, p.835.

5. Thomas Berry, telephone conversation, 2004; and Introduction to Tom Hayden (1996) *The Lost Gospel of The Earth*, Sierra Club Books, San Francisco, p.xv.

6. *The Homeric Hymns*, trans. Jules Cashford (2003), Penguin Classics, London, pp.140–41.

7. Hesiod, *Theogony*, trans. Dorothea Wender, in *Hesiod and Theognis* (1973) Penguin Classics, London, pp.27–29.

8. Aeschylus, *The Libation Bearers*, in *Greek Tragedies*, vol. ii, trans. David Grene and Richmond Lattimore, University of Chicago Press, Chicago and London, p.7.

9. Sophocles, *Oedipus Rex*, in *Greek Tragedies*, vol.i, trans. Grene and Lattimore, p.112.

10. I am indebted to Satish Kumar for all the references to Indian thought.

11. Lucius Apuleius (1950) *The Golden Ass*, trans. Robert Graves, Penguin Books, London, p.228.

12. Joseph Campbell, *The Inner Reaches of Outer Space: Metaphor as Myth and Religion*, (1986) New York, Alfred van der Marck Editions and Toronto, St James's Press Ltd, *passim*.

13. Stephan Harding (2009) *Animate Earth: Science, Intuition and Gaia*, Green Books, Totnes.

14. Jung, *The Tavistock Lectures*, p.682.

15. Logion 77. *The Gospel According To Thomas*, Coptic Text established and translated by A. Guillaumont *et al.* (1976) E.J. Brill, Leiden.

16. Albert Einstein, *The Expanded Quotable Einstein*, ed. Alice Calaprice (2000), Princeton University Press, Princeton and Oxford, p.316.

17. See Jules Cashford (2003) *The Moon: Myth and Image*, Cassell Illustrated, London.

5. In Service to Gaia

JAMES LOVELOCK

In 1925 the American scientist Alfred Lotka published a small but important book, *Physical Biology*. In it he wrote: 'It is not so much the organism or the species that evolves, but the entire system, species and environment. The two are inseparable'. Lotka's view of evolution passed almost unnoticed in his time and it was not until the National Aeronautics and Space Administration (NASA) in the 1960s began exploring our planetary neighbourhood that this broader, view of the Earth was revisited. As part of NASA's exploration team, it led me to propose, in a paper in *Nature* in 1965, that life and its environment are so closely coupled that the presence of life on a planet could be detected merely by analysing chemically the composition of its atmosphere. This proposal is now part of NASA's astrobiology programme and they aim to use it in the search for life on extra-solar planets.

When we look at the Earth we see an atmosphere that, apart from the noble gases, has a composition almost wholly determined by the organisms at the surface. If some catastrophe removed all life from the Earth without changing anything else, the atmosphere and surface chemistry would rapidly – in geological terms – move to a state similar to those of Mars or Venus. These are dry planets with atmospheres dominated by carbon dioxide and close to the chemical equilibrium state. By contrast, we have a cool wet planet with an unstable atmosphere that stays constant and always fit for life. The odds against this are close to infinity.

Science is about probabilities, so we are forced to consider the difficult but more probable alternative: something regulates the atmosphere. What is it? It has to be something connected with life at the surface, because we know that the atmospheric gases, oxygen, methane and nitrous oxide, are almost wholly biological products, while others, nitrogen and carbon dioxide, have been massively changed in abundance by organisms.

Moreover, the climate depends on atmospheric composition and there is evidence that the Earth has kept a fairly comfortable climate ever since life began, in spite of a 30% increase in solar luminosity. Together these facts led me to propose, in a 1969 paper in the JAAS *(Journal of the American Astronautical Society),* that the biosphere was regulating the atmosphere in its own interests. Two years later I started collaborating with Lynn Margulis and we published a paper in *Tellus* where we stated:

> The Gaia hypothesis views the biosphere as an active adaptive control system able to maintain the Earth in homeostasis.

In 1981 I redrafted the hypothesis as an evolutionary model, 'Daisyworld', that was intended to show that self-regulation can take place on a planet where organisms evolve by natural selection in a responsive environment. Following the model, the Gaia hypothesis was restated as follows:

> The evolution of organisms and their material environment proceeds as a single tight-coupled process from which self-regulation of the environment at a habitable state appears as an emergent phenomenon.

Large ideas in science take about forty years to be accepted. They have to be fought over and tested and preferably expressed also in mathematics. As you probably know, Gaia Theory has had some fierce battles, especially with biologists such as Richard Dawkins. They were painful at the time but I am grateful to my opponents, for they sharpened the theory. They were right to object to my idea that life regulates the Earth, and their criticism made me realise that the regulation came from the whole system, life and the material environment, tightly coupled as the single entity, Gaia. The theory is now widely accepted as describing the Earth. In 2001, at a huge meeting in Amsterdam of all branches of science, more than a thousand delegates signed a declaration that stated as its first point: 'The Earth is a self-regulating system made up from all life, including humans, and from the oceans, the atmosphere and the surface rocks'.

At last Gaia Theory was part of science but, sadly, it was about twenty years too late. If we had accepted in, say, 1980 that the Earth was in effect alive, we would have known that we cannot pollute the air and use the

Earth's skin – its forest ecosystems – as a mere source of products to feed ourselves and furnish our homes. Those ecosystems, before we destroyed them, were regulating the climate and atmospheric composition. Our planet is now at a state that in medicine would be called 'failure'. It is at a crisis point beyond which it will soon move to a different stable state where it can more easily maintain itself physiologically. Gaia is normally healthy but in its three-and-a-half-billion-year life it has suffered from fevers several times before – the last was 55 million years ago at the beginning of the period the geologists call the Eocene.

At that time an accidental geological event released into the air between 1 and 2 million million tons of carbon dioxide. We are fairly sure about this from measurements made by Henry Elderfield of Cambridge University and his colleagues. They measured the carbon and oxygen isotopes of the sedimentary rocks of that time and confirmed the quantity of carbon put into the air and the extent the temperature changed. Putting this much CO^2 into the air caused the temperature of the temperate and Arctic regions to rise by 8°C and of the tropics by about 5°C and it took about 200,000 years for conditions to return to the state before the fever. In the twentieth century we released by burning fossil fuel about half that amount of CO^2 and if we continue as we are, we also will have released in thirty years from now more than a million million tons of CO^2. Moreover, the Sun is now hotter than it was 55 million years ago and we have disabled about 40% of Gaia's regulatory capacity by using land to feed people and animals. This is why climate scientists are so concerned that we have caused irreversible climate change.

With the Earth's climate in net positive feedback it is not surprising that global warming is turning out to be much more serious than we thought only a few years ago. Systems in positive feedback act as amplifiers and any heating effect will be greater than predicted by classical climatology; and there is the danger of instability, something that will lead to surprises – events more deadly than we had imagined. There has been one already: the summer of 2003 in Europe, when more than thirty thousand people died from the heat.

The signs are that we do not have long to act before global warming starts the irreversible move to much higher temperatures. In a BBC Horizon programme a few years ago, Peter Cox of the Hadley Centre introduced the concept of global dimming. He presented evidence to suggest that the widespread haze of smog from cars and industry covered the Northern hemisphere and largely offset the global warming. The residence time of smog particles is only a few days, whereas that of carbon dioxide is about one hundred years. Any economic downturn or planned cutback in fossil-fuel use that lessened the smog could carry us beyond the threshold of irreversible change. If it does there will be vast geographic as well as climatic changes. In many ways we live in a fool's climate. We are damned if we do and damned if we don't.

In no way do I mean that there is no hope for us or that there is nothing that we can do. I see our predicament as like that of the UK in 1940 when it was about to be invaded by a powerful enemy; now we are at war with the Earth and are faced with much more than a *Blitzkrieg*. We will do our best to avoid a catastrophe, but sadly in the present world the green concepts of sustainable development and renewable energy that inspired the Kyoto agreement are too late. They might have worked fifty years ago but now they are false and beguiling dreams that can only lead to failure. I cannot see the United States, or the emerging economies of China and India, cutting back in time. I fear that the worst may happen and our survivors will have to adapt to a hot and uncomfortable world. To retain civilisation they will need more than ever a secure and reliable source of energy to power the adaptation. Any large city would die in a week without electricity; all the services we take for granted depend on it, and this is why we need the security of a powered descent – and for that I believe that there is no sensible alternative to nuclear energy.

Long overdue is a change in the way we think about the Earth. We have to stop thinking that we are in charge and are the stewards of the Earth. We have to abandon the idea that the only thing that matters is the welfare of humankind and that the Earth was given to us for our benefit alone. We have for too long behaved badly towards the host of other life forms with which we share our planet and on whom we depend for a habitable environment. Although we pay lipservice to threats to wildlife and to ecosystems like coral reefs, the Amazon forests, and the fresh-water ecosystems here in the UK, we are in practice obsessed with environmental hazards to personal health such as pesticide residues in foodstuff, nuclear radiation, and genetically modified food. And to judge from the supermarket shelves, we seem to have the illusion that if the whole planet were farmed organically all would be well. To the contrary: I think it extremely unlikely that the regulatory functions of natural ecosystems can simply be replaced by farmland.

Perhaps the saddest thing is that if we fail, Gaia will lose as much as or more than we do. Not only will wildlife and whole ecosystems become extinct, but in human civilisation the planet has a precious resource. We are not merely a disease: we are through our intelligence and communication the nervous system of the planet.

Good science though it is, perhaps the most useful contribution from Gaia is in the public domain. Organisms, and this includes people, that improve their environment make it better for their progeny, whereas those that foul the environment worsen their chances. Moreover, the self-regulation of Gaia requires the existence of firm constraints. This gives Gaia an ethical significance. We are truly accountable to the Earth. Most of us need more than an accurate explanation of Life and the Cosmos. We need in addition something to revere and respect. The Earth is the beautiful anomaly of the solar system; it is tangible and could fill this need. Names are important and Gaia is a good name for the system. People do not revere portmanteau words or phrases like geobiochemistry or geosphere-biosphere systems but they can and do see the word Gaia embracing both the intuitive side of science and the wholly rational understanding that comes from Earth System Science. Think of Gaia as the name of the Earth, a name that makes it a personal presence for all of us .

Through us Gaia has seen herself from space and seen how beautiful she is, and she begins to know her place in the universe. We should be the heart and mind of the Earth, not its malady. So let us be brave and cease thinking of human needs and rights alone and see that we have harmed the living Earth and need to make our peace with Gaia. Most of all, we should remember that we are a part of Gaia and it is indeed our home.

References

Harding, S.P. & Margulis, Lynn (2010) 'Water Gaia: Three and a Half Thousand Million Years of Wetness on Planet Earth.' In: *Gaia in Turmoil,* MIT Press.

Lovelock, James E. (1995) *The Ages of Gaia,* Oxford University Press

—, (2000) *Gaia: The Practical Science of Planetary Medicine,* Gaia Books, UK.

—, (2006) *The Revenge of Gaia.* Penguin, Allan Lane, UK.

—, (2009) *The Vanishing Face of Gaia.* Penguin, Allan Lane, UK.

Margulis, Lynn & Sagan, Dorion (1997) *Microcosmos,* University of California Press.

6. Systems Thinking and the State of the World: Knowing How to Connect the Dots

FRITJOF CAPRA

As our new century unfolds, it is becoming more and more evident that the major problems of our time – energy, the environment, climate change, food security, financial security – cannot be understood in isolation. They are systemic problems, which means that they are problems of an entire system as a whole, and therefore all interconnected and interdependent. One of the most detailed and masterful documentations of the fundamental interconnectedness of world problems is the new book by Lester Brown, *Plan B*.[1] Lester Brown, founder of the Worldwatch Institute and, more recently, of the World Policy Institute, has been for many years one of the most authoritative environmental thinkers. In this book, he demonstrates with impeccable clarity how the vicious circle of demographic pressure and poverty leads to the depletion of resources – falling water tables, shrinking forests, collapsing fisheries, eroding soils, and so on – and how this resource depletion, exacerbated by climate change, produces failing states whose governments can no longer provide security for their citizens, some of whom in sheer desperation turn to terrorism.

All these problems, ultimately, must be seen as just different facets of one single crisis, which is largely a crisis of perception. It derives from the fact that most people in our society, and especially our large social institutions, subscribe to the concepts of an outdated worldview, a perception of reality inadequate for dealing with our overpopulated, globally interconnected world.

Plan B is Lester Brown's roadmap for saving civilisation. It is the alternative to 'business as usual' (Plan A), which leads to disaster. The main message of the book is that there *are* solutions to the major problems of our time; some of them even simple. But they require a radical shift in our perceptions, our thinking, our values. And, indeed, we are now at the beginning of such a fundamental change of worldview in science and society, a change of paradigms as radical as the Copernican Revolution.[2]

Unfortunately, this realisation has not yet dawned on most of our political leaders. The recognition that a profound change of perception and thinking is needed if we are to survive has not yet reached our corporate leaders either, nor the administrators and professors of our large universities.

Most of our leaders are unable to 'connect the dots', to use a popular phrase; they fail to see how the major problems of our time are all interrelated. Moreover, they refuse to recognise how their so-called solutions affect future generations. From the systemic point of view, the only viable solutions are those that are sustainable. As Lester Brown put it in his pioneering definition more than twenty-five years ago, 'A sustainable society is one that satisfies its needs without diminishing the prospects of future generations'.[3]

Systems thinking

Over the past twenty-five years it has become clear that a full understanding of these issues requires nothing less than a radically new conception of life. And indeed, such a new understanding of life is now emerging.[4] At the forefront of contemporary science, the universe is not longer seen as a machine composed of elementary building blocks. We have discovered that the material world, ultimately, is a network of inseparable patterns of relationships; that the planet as a whole is a living, self-regulating system. The view of the human body as a machine and of the mind as a separate entity is being replaced by one that sees not only the brain, but also the immune system, the bodily tissues, and even each cell as a living, cognitive system. Evolution is no longer seen as a competitive struggle for existence, but rather as a cooperative dance in which creativity and the constant emergence of novelty are the driving forces. And with the new emphasis on complexity, networks, and patterns of organisation, a new science of qualities is slowly emerging.

This new conception of life involves a new kind of thinking – thinking in terms of relationships, patterns, and context. In science, this way of thinking is known as 'systemic thinking', or 'systems thinking'. It emerged in the 1920s and 1930s from a series of interdisciplinary dialogues among biologists, psychologists, and ecologists.[5] In all these fields, scientists realised that a living system – an organism, ecosystem, or social system – is an integrated whole whose properties cannot be reduced to those of smaller parts. The 'systemic' properties are properties of the whole, which none of its parts have. So, systems thinking involves a shift of perspective from the parts to the whole. The early systems thinkers coined the phrase: 'The whole is more than the sum of its parts'.

What exactly does this mean? In what sense is the whole more than the sum of its parts? The answer is: relationships. All the essential properties of a living system depend on the relationships among the system's components. Systems thinking means thinking in terms of relationships. Understanding life requires a shift of perspective, not only from the parts to the whole but also from objects to relationships. These relationships include the relationships among the system's components and also those between the system as a whole and surrounding larger systems. Those relationships between the system and its environment are what we mean by context. Systems thinking is always contextual thinking.

Understanding relationships is not easy for us, because it is something that goes counter to the traditional scientific enterprise in Western culture. In science, we have been told, things need to be measured and weighed. But relationships cannot be measured and weighed; relationships need to be mapped. So there is another shift of perspective: from measuring to mapping, from quantity to quality.

When we map relationships, we find certain configurations that occur repeatedly. This is what is called a pattern. Networks, cycles, feedback loops, are examples of patterns of organisation that are characteristic of life. The outstanding property these patterns have in common is that they are nonlinear. All living systems are nonlinear systems. Hence, systems thinking also involves a shift from linear to nonlinear thinking.

In science, the study of nonlinear systems is exceedingly difficult, and until recently the corresponding nonlinear equations were impossible to solve. But in the 1970s scientists for the first time had powerful high-speed computers that could help them tackle and solve nonlinear equations. In doing so, they devised a number of techniques, a new kind of mathematical language, known technically as 'nonlinear dynamics' and popularly as 'complexity theory', that revealed very surprising patterns underneath the seemingly chaotic behaviour of nonlinear systems.[6]

When you solve a nonlinear equation with these new techniques, the result is not a formula but a visual shape, a pattern traced by the computer.

So, this new mathematics is a mathematics of patterns, of relationships. The strange attractors of chaos theory and the fractals of fractal geometry are examples of such patterns. They are visual descriptions of the system's complex dynamics.

Ecological literacy

When we apply the new conception of life to studying the structures, metabolic processes, and evolution of the myriads of species on the planet, we notice immediately that the outstanding characteristic of our biosphere is that it has sustained life for billions of years. How does the Earth do that?

To understand how nature sustains life, we need to move from biology to ecology, because sustained life is a property of an ecosystem rather than a single organism or species. Over billions of years of evolution, the Earth's ecosystems have evolved certain principles of organisation to sustain the web of life. Knowledge of these principles of organisation, or principles of ecology is what is meant by 'ecological literacy', or 'ecoliteracy'.[7]

In the coming decades the survival of humanity will depend on our ecological literacy – our ability to understand the basic principles of ecology and to live accordingly. This means that ecoliteracy must become a critical skill for politicians, business leaders, and professionals in all spheres, and should be the most important part of education at all levels – from primary and secondary schools to colleges, universities, and the continuing education and training of professionals.

We need to teach our children, our students, and our corporate and political leaders, the fundamental facts of life – that one species' waste is another species' food; that matter cycles continually through the web of life; that the energy driving the ecological cycles flows from the sun; that diversity assures resilience; that life, from its beginning more than three billion years ago, did not take over the planet by combat but by networking.

All these principles of ecology are closely interrelated. They are just different aspects of a single fundamental pattern of organisation that has enabled nature to sustain life for billions of years. In a nutshell: nature sustains life by creating and nurturing communities. No individual organism can exist in isolation. Animals depend on the photosynthesis of plants for their energy needs; plants depend on the carbon dioxide produced by animals, as well as on the nitrogen fixed by bacteria at their roots; and together plants, animals, and microorganisms regulate the entire biosphere and maintain the conditions conducive to life.

Sustainability, then, is not an individual property but a property of an entire web of relationships. It always involves a whole community. This is the profound lesson we need to learn from nature. The way to sustain life is to build and nurture community. A sustainable human community interacts with other communities – human and nonhuman – in ways that enable them to live and develop according to their nature. Sustainability does not mean that things do not change. It is a dynamic process of coevolution rather than a static state.

Interconnectedness of world problems

Once we become ecologically literate, once we understand the processes and patterns of relationships that enable ecosystems to sustain life, we will also understand the many ways in which our human civilisation, especially since the Industrial Revolution, has ignored these ecological patterns and processes and has interfered with them. And we will realise that these interferences are the fundamental causes of many of our current world problems. Thinking systemically, we will recognise the major problems of our time as systemic problems – all interconnected and interdependent. This is the fundamental message of the first part of Lester Brown's book, *Plan B*, in which he offers a detailed systemic analysis to document the fundamental interconnectedness of our current world problems.

This part is predictably depressing, but the second part is optimistic and exciting.

The strategy of Plan B is informed by the awareness of the systemic interdependence of the major problems of our time. It involves several simultaneous actions that are mutually supportive, mirroring the interdependence of the problems they address.

The theoretical framework underlying Plan B is based on the thorough understanding of the basic principles of ecology. Its detailed proposals involve applying this ecological knowledge to the redesign of our technologies and social institutions, so as to bridge the current gap between human design and the ecologically sustainable systems of nature. This is what is known as ecological design, or 'ecodesign'. In recent years, there has been a dramatic rise in ecologically oriented design practices and projects, all of which are now well documented.[8]

These ecodesign technologies and projects all incorporate the basic principles of ecology and therefore have some key characteristics in common. They tend to be small-scale projects with plenty of diversity, energy efficient, non-polluting, community oriented, and labour intensive, creating plenty of jobs.

The recent ecodesign revolution provides the justification for Lester Brown's optimism and hope. All the proposals of Plan B are based on existing technologies and illustrated with successful examples from countries around the world. They make it evident that we have the knowledge, the technologies, and the financial means to save civilisation and build a sustainable future. What we need now is political will and leadership.

Notes

1. Lester Brown (2008) *Plan B 3.0*, Norton, New York.
2. See Fritjof Capra (1982) *The Turning Point*, Simon & Schuster.
3. Lester Brown (1981) *Building a Sustainable Society*, Norton, New York.
4. See Fritjof Capra (2002) *The Hidden Connections*, Doubleday, New York.
5. See Fritjof Capra (1996) *The Web of Life*, Anchor Books, pp.15ff.
6. See Capra (1996) pp.112ff.
7. *Ibid.*, pp. 297ff.
8. See Capra (2002) pp.233ff.

7. Exploration and Theory in Science

CRAIG HOLDREGE

Western Science has given us many benefits, but it has also helped to disconnect us from the world by focusing on theory and abstraction. If science is to help us to resolve the ecological crisis it will have to include ways of knowing that so far it has either neglected or actively refused to seriously consider. Johann Wolfgang von Goethe (1749–1832) developed an exploratory, phenomenological approach to science that has much to offer us in the effort of creating an expanded science that integrates careful observation and critical thinking with intuitive understanding – a kind of science that could help us to develop a truly beneficial relationship with nature.

So what can we say about Goethe's approach to science? In a discussion of Goethe's research into colour and light, Ribe and Steinle (2002) characterise his approach as 'exploratory' and contrast it with the more common 'theory-oriented' approach in mainstream science. It is this contrast that I want to explore in this chapter, first by focusing on Goethe's approach and then by discussing the interplay of exploration and theory in Darwin's work.

Goethe's delicate empiricism

In his extensive research into light and colour, Goethe performed countless experiments, continually varying the conditions in new ways (Goethe 1995, Chap. 7). Even a critic of Goethe's approach would have to grudgingly admire the persistence with which he returned again and again to observation and never tired of looking at phenomena from new angles. Why did he do this? First, he recognised that one can speak about light and colour only in relation to the conditions under which they appear. If we are going to do justice to phenomena we have to look at them within a variety of contexts. Secondly, he was wary of our human propensity to form judgments about things based upon few observations or experiments. As he writes in his seminal essay, written in 1792, 'The Experiment as Mediator of Object and Subject':

> We cannot take great enough care when making inferences based on experiments. We should not try through experiments to directly prove something or to confirm a theory. For at this pass – the transition from experience to judgment, from knowledge to application – lie in wait all our inner enemies: imaginative powers that lift us on their wings into heights while letting us believe we have our feet firmly on the ground, impatience, haste, self-satisfaction, rigidity, thought forms, preconceived opinions, lassitude, frivolity, and fickleness. This horde and all its followers lie in ambush and suddenly attack both the active observer and the quiet one who seems so well secured against all passions.
> (2010, p.20)

Faithfulness to the phenomena keeps our judgments grounded and connected to reality. We become cautious and circumspect and do not get carried away by the levity of a grand idea. Goethe came to the conclusion that 'one experiment, and even several of them, does not prove anything and that nothing is more dangerous than wanting to prove a thesis directly by means of an experiment' (2010, p.20). By varying the conditions of experiments and by looking at things from different angles, we gain a comprehensive and differentiated picture that helps us leave behind all-too-narrow, schematic, or rigid conceptions. Therefore, in Goethe's view, 'we accomplish most when we never tire in exploring and working through a single experience or experiment by investigating it from all sides and in all its modifications' (2010, p.22).

Goethe was motivated by his love of the phenomena and his desire to do justice to them in scientific explorations. Scientists should work to 'take the measure for knowledge – the data that form the basis for judgment – not out of themselves but out of the circle of what they observe' (2010, p.19). Toward the end of his life Goethe called this approach 'a delicate empiricism that makes itself utterly identical with the object' (1995, p.307). It is empiricism, because it orients itself always around the way phenomena actually appear and explores them in their manifold variations. It is delicate, because it does not want to subjugate the phenomena to a particular theory or a general idea as does mainstream science. Goethe felt the tendency in himself and others to overpower the phenomena with ideas.

We become delicate empiricists when our primary goal is to let the phenomena speak for themselves and therefore explore them from many sides while remaining highly conscious about the way we interact with the phenomena through our ideas. I suggest that this is a way for scientific ideas to become more in tune with nature so that we can better address environmental issues.

Theory as beholding

Within mainstream science a theory or hypothesis is understood as an idea that the scientist comes up with to explain the phenomena in question. 'Scientific hypotheses and theories are not *derived* from observed facts, but *invented* in order to account for them' (Hempel 1966, p.15; his emphasis). A theory is a specific conceptual framework that orders and makes intelligible a variety of phenomena. Theories and hypotheses guide research and help in the formulation of research questions; they are the lenses that help to focus an investigation. The more all-encompassing a theory is, the more abstract it is. As an abstraction it remains distinct from the phenomena.

For physicists since Galileo and Newton, the highest form of theoretical explanation is a mathematical equation that allows one to predict the behaviour of phenomena. The best explanations should be simple, elegant, logical and, ideally, universal (Wolpert 1994).

Goethe was not interested in creating theories in this modern sense of the word. Zajonc writes that 'in common with all scientists, Goethe searched for patterns, the hidden lawfulness within the welter of phenomena, but for him they were to be exalted perceptual experiences, not abstract substitutes for nature's glory' (1993, p.203). For Goethe, important scientific ideas could be more than general abstractions, a notion that does not sit well with the modern scientific mind.

He spoke of 'experiences of a higher kind' that emerge out of the careful and extensive exploration of phenomena (2010, p.22). The phenomena illuminate each other and this illumination increases as the breadth and depth of our experience grows until it reaches such clarity that we can speak of an idea expressing itself through the phenomena.

Or I could also say: the ideal aspect of the phenomenon, which makes it intelligible, comes to appearance as the 'illuminated phenomenon'.

Goethe also called such experiences of a higher nature 'true theory': 'There is a delicate empiricism that makes itself utterly identical with the object, thereby becoming true theory' (1995, p.307). In this remarkable use of the term 'theory', Goethe re-awakens the way theory was understood in Greek philosophy. As Barfield says (1988), theory for Aristotle is 'the moment of fully conscious participation, in which the soul's *potential* knowledge (its ordinary state) becomes *actual*, so that man can at last claim to be awake' (p.49). Theory in this sense is the wakeful perceiving of the mind; it is a beholding. Or in Gadamer's words, theory means '"being present" in the lovely double sense that means that the person is not only present, but completely present' (Gadamer 1998, p.31).

The distinction between an abstract idea and an idea that arises in the encounter with the phenomena is essential if we are to develop a science that is attuned to the phenomena of the world (Bortoft 1996; Maier, Brady, & Edelglass 2008). Unfortunately, interpreters of Goethe's approach often miss this essential point (e.g., Richards 2002; Tantillo 2002). They don't see that Goethe's approach leads to an experience in which the phenomenon discloses itself as a sensory-ideal reality. Ernst Cassirer captures the essential distinction when he writes, 'the mathematical formula aims to explain the appearances. The aim of Goethe's approach is to make the appearances fully visible' (1971, p.78; translation CH).

So there is a clear distinction between the theory-driven, explanatory approach in science and the Goethean approach of delicate empiricism. In a theory-driven approach the human-created theory becomes a primary context for the research, while in Goethe's approach the encounter with the phenomena provides the primary context and as a result the abstract notion needs to be overcome. Within a Goethean perspective, if we use the term theory at all, it is to point to a certain culmination of the research process in which there is a direct seeing-beholding of the phenomenal relations themselves.

A dynamic tension

As clear as the difference in approaches may be, there is also a danger in any dichotomisation that paints a black-and-white, either-or picture. Reality is more complex, more fluid, more messy. On the one hand, when scientists work within a conventional hypothesis- or theory-oriented approach, they are of course always being nudged by the phenomena, which include experimental results, to expand or even discard their previous notions. Those who make new discoveries are especially open to what the phenomena tell them. On the other hand, any one who works in a more exploratory fashion is going to be guided at least by questions or hunches – and these are idea-filled. As points of view, ideas provide the initial context for our interaction with the phenomena. The important thing is that we are conscious of such ideas, that we hold them flexibly, and that we let them be modified by new experiences.

In all scientific inquiry there is a tension between exploration – the search for new understanding and the uncovering of new phenomena – and the desire to understand, to find patterns and lawfulness in the world. As Goethe writes:

> We take pleasure in a thing in so far as we form an idea of it and when it fits into our way of looking. We may try to raise our mode of thought so far above the everyday mode as possible and strive to purify it, but nonetheless it usually still remains only a mode of thought. It follows that we attempt to bring many phenomena into a certain graspable relation to one another that they may, looked at more closely, not have. And we have the tendency to form hypotheses and theories and to craft terminology and systems accordingly. We cannot condemn these attempts since they arise with necessity out of the organization of our being.
> (2010, p.20)

Nonetheless, we can work to move beyond such limitations. Delicate empiricism is a method to overcome the inertia of the human mind to remain within already formed paradigms and well-trodden mental pathways. To transcend deep-seated habits of thought in the search for more adequate insights into the phenomena requires an arduous transformation of human capacities – and this is a lifelong (at least) process. For this reason Goethe added to his words about delicate empiricism that such 'enhancement of our mental powers belongs to a highly evolved age' (1995, p.307). Along the way we can make use of hypotheses and conceptual frameworks. What is important is how we use them (Zajonc 1999).

Goethe is critical of hypotheses because they constrict: 'All hypotheses hinder us in beholding and considering the objects and phenomena from all sides' (1973, p.441). Therefore it is important to gain a free and sovereign relation to one's hypotheses: 'Hypotheses are the scaffolding that one erects in front of a building and that are dismantled when the building is completed. To the worker the scaffolding is indispensable, but he must not take it for the building itself' (p.441). As Goethe remarks, it is a freeing experience to cast off a way of seeing and to venture into a new one:

> When we free our mind from a hypothesis that unnecessarily limited us, that compelled us to see falsely or half-way and to combine falsely, that compelled us to brood rather than to look, to commit sophistry rather than to discern, then we have performed a great service. For then we can see the phenomena freely and in other relations and connections; we can order them in our own way and have occasion to err in our own way – an invaluable opportunity as long as we are able to understand our own errors.
> (Goethe 1973, p.441)

What is key is that the scientist remains at all times aware of how her concepts are at work in the process of coming to understand. All good science needs grounding, therefore, in what Bentz and Shapiro (1998) call 'mindful inquiry', a process in which the scientist is 'aware of the epistemological grounds on which his inquiry rests' (p.32). Without this awareness we always run the danger of conflating our ideas with the phenomena themselves, not realising that we are speaking of a phenomenon only in the restricted terms of a specific hypothesis, framework or theory. This is especially crucial as a healthy deterrent to the widespread tendency to make universal claims, which is essentially a form of unwarranted extrapolation and reveals lack of context sensitivity as well as a mind all too confident in its powers of generalisation. The question is not whether we should have ideas that guide our investigations. Rather, the question is: What is the quality of those ideas? Do we apply them consciously and use them as ways of illuminating the phenomena from a certain perspective? Can we disengage ourselves from a theory or hypothesis so that the phenomena can reveal new sides and dimensions? Are we ready to let the encounter with the phenomena be the stimulus for our ideas to grow and transform, and maybe even then to fall away? The vitality and depth of a scientific idea would show itself in the way it grows and transforms in the course of the dialogue with the phenomena.

Darwin: exploration and theory

The work of Charles Darwin highlights in a vivid way the relation between exploration and theory formation in science. As a young man Charles Darwin was an avid naturalist (Desmond & Moore 1994). He loved to observe and collect. He was able to live out this passion when at the age of twenty-two he became the naturalist for the *Beagle*, a ship that sailed around the globe over the course of five years (1831–36). During the journey on the *Beagle* he collected many fossils and began to build up a picture of how landscapes were formed through geological time. He observed and collected plants and animals in different environments and noticed how they were exquisitely adapted to the conditions in which they lived. He had an eye for the subtle differences between specimens of a species that could easily be overlooked. He saw, wherever he looked, great variation in nature and, in the years that followed his return to England, he came to the conclusion that that there are no fixed boundaries between species. One could say that, guided by the immersion in the phenomena themselves, Darwin began to form an idea of the transformational nature of life on earth – not only within the individual but also within populations, species, and whole landscapes. For Darwin, nature was characterised by variation, change, and flux, and having gained this view, he saw himself leaving behind the more static and dominant view of his time that species are separate entities created by God. The young Darwin was working in an exploratory Goethean spirit and contributing to a living, dynamic, and ecologically informed view of life on earth.

Throughout his long life Darwin never tired of observing and pondering the significance of natural phenomena, whether it was the contribution of earthworms to soil formation, the movements of plants, or the esoteric biology of barnacles. But Darwin was not satisfied to build up a dynamic picture of life on earth. He wanted to explain how organisms evolved. What did 'explain' mean to him? It meant finding a general theory – a causal mechanism – that could show how species transform and become so well adapted to their environments. He was compelled by the quest to find one general, fundamental and logical idea – an urge that is so characteristic of modern science and that culminates in the desire to find the one theory of everything. During his life Darwin's exploratory manner of research became increasingly overlaid by a theory-driven one.

In formulating his theory of evolution through natural selection Darwin was strongly influenced by Thomas Malthus' *Essay on the Principle of Population* (written in 1798). Malthus argued that human population has a tendency to grow that is 'indefinitely greater than the power in the earth to produce subsistence for man' (1999, p.13). As a result, illness, wars, famines, and natural disasters become important factors in controlling population growth. In his seminal book, *Origin of Species* (first published in 1859), Darwin wrote that his theory is 'the doctrine of Malthus applied with manifold force to the whole animal and vegetable kingdoms' (1979, p.117). This is a classic statement of a theory-driven approach: one finds an idea, generalises it and then applies it far beyond the boundaries of the phenomena within which it was originally conceived.

It was important for Darwin that the basic tenets of his theory seemed to follow from facts of nature. Since he had so much knowledge of the natural world, he was well aware of the immense challenge of finding a theory that did, to his mind, do justice to the phenomena. He brought together in his mind three areas of phenomena that were based on thorough observation: There is heritable variation within species and populations; there are many more offspring produced by plants and animals than actually survive; plants and animals are well adapted to their environments. He then connected these phenomena in thought: Since not all offspring survive, it makes sense that those that do survive will be those variations that are better adapted to the environment. He spoke of natural selection as the 'principle by which each slight variation, if useful, is preserved' (1979, p.115). When well-adapted variations produce fertile offspring, they become the dominant form of the species. With enough time, the species will evolve provided that new variations are being produced that are better-adapted to existing conditions or that are well-adapted to a changed environment. Evolution occurs as the additive effect of many small changes in organisms resulting from the 'struggle for existence' (1979, Chapter 3) over long periods of time. This is a compelling train of thought and once it became clear in his mind, Darwin proceeded to interpret all biological phenomena in its light.

Remarkably, with a keen eye for his own thought processes, near the end of his life Darwin noted how his mind had changed as a result of his exclusive concern with the theory of evolution through natural selection. Looking back to what he had written in his journal during the *Beagle* voyage, he remarked:

> In my Journal I wrote that whilst standing in the midst of the grandeur of a Brazilian forest, 'it is not possible to give an adequate idea of the higher feelings of wonder, admiration, and devotion which fill and elevate the mind'. I well remember my conviction that there is more in man than the mere breath of his body. But now the grandest scenes would not cause any such convictions and feelings to rise in my mind. It may be truly said that I am like a man who has become colour-blind, and the universal belief by men of the existence of redness makes my present loss of perception of not the least value as evidence.
> (Darwin 2005, p.76)

In other words, his worldview had taken on firm contours and in the wake of the development of his theory, the feelings of wonder and reverence he previously felt had dissipated. He could see the natural world only through the lens of his theory and believed that what he saw through this lens was true, and more importantly, it was all that was needed to explain nature – nature contains nothing or is expressive of nothing beyond that which his theory encompasses. For this reason, he sees the loss of his earlier feelings of wonder and devotion as the loss of delusory feelings toward something higher or deeper in nature – the divine or God. Everyone else in the world may claim that he is colour blind, but that in no way sways him. Darwin observed that 'my mind seems to have become a kind of machine for grinding general laws out of large collections of facts' (2005, p.113). And in the end he saw these general laws as being the laws that govern the natural world.

Today Darwin's theory (of course modified and refined by over a century of further scientific work) has become, in Thomas Kuhn's sense 'normal science' (Kuhn 1996, Chapter 3). It is the accepted paradigm. All biology students learn to interpret phenomena through the lens of evolution through natural selection. Whatever biological characteristic one confronts, the question is essentially the same: How does this characteristic contribute to the survival of the organism? If I can discover how it increases or at least does not hinder the fitness of the organism, I believe I have understood it. It is quite easy to come up with theoretical claims about how it is beneficial for a bird to have just the wing structure it does, how the highly developed forelegs of a mole allow it to live the life of burrowing animal, or how the long neck of the giraffe lets it feed on leaves higher up in trees than are accessible to most other four-legged animals. After all, animals and plants that exist are well-adapted and all their characteristics do contribute in one way or another to their survival.

But it is another matter to try to explain the evolutionary process – to formulate the set of causal relations – that brought forth such adaptations. For example, when applying his theory to giraffe evolution Darwin writes (in the 1872 sixth edition of *Origin of Species*):

> The giraffe, by its lofty stature, much elongated neck, fore-legs, head and tongue, has its whole frame beautifully adapted for browsing on the higher branches of trees. It can thus obtain food beyond the reach of the other Ungulata or hoofed animals inhabiting the same country; and this must be a great advantage to it during dearths ... Those individuals which had some one part or several parts of their bodies rather more elongated than usual, would generally have survived. These will have intercrossed and left offspring ... whilst the individuals, less favoured in the same respects will have been the most liable to perish ... By this process long-continued ... it seems to me almost certain that an ordinary hoofed quadruped might be converted into a giraffe. (pp.177ff.)

With this description Darwin paints a picture of how the giraffe could have evolved its long neck. It makes sense and seems logical. It is a typical example of how characteristics are explained within a Darwinian framework. But this kind of explanation is actually, as Gould and Lewontin (1979) call it, an 'adaptive story'. It may or may not be true; there may be alternative stories that are just as logical and compelling. For example, biologists have come up with other adaptive stories that 'explain' the giraffe's long neck, such as competition between males and thermo-regulation (see Holdrege 2005). I think you would be hard pressed to find any characteristic for which one could not find at least two different compelling adaptive stories. As Gould and Lewontin (1979) point out, it is all too easy to fall into the trap of confusing 'the fact that a structure is used in some way with the primary evolutionary reason for its existence and conformation' (p.587). Because a long-necked giraffe does occasionally reach high into trees to feed (usually at times in which there is adequate food supply!), does not mean this is the evolutionary reason for its existence. Such conflation is a form of self-forgetfulness that the theoretical mind so easily succumbs to in its search for one-size-fits-all explanations.

The more you study an animal – or any other organism – in detail, the less convincing adaptive stories become. They pale before the richness of the whole organism itself. It is fascinating to follow how Darwin unfolded his theory and to read how he tries to apply it to phenomenon after phenomenon. In the process, the abstract notion, once formed and consolidated, tends to take on a life of its own – revealing everywhere the same 'explanation' only because it adamantly remains true to itself and takes into account just those aspects of the phenomena that fit into the framework. This is when theory runs the danger of becoming the kind of 'lethal generality' Goethe so hoped science based on a delicate empiricism would avoid (1995, p.61).

For example, Darwin writes, 'nature cares nothing for appearances, except in so far as they may be useful to any being' (p.132). This follows from his basic premise; if his theory is right, then this must be the case. But this exclusively utilitarian point of view also radically limits the way one is 'allowed' to judge organic forms. When you free yourself from this bias, enter a more exploratory mode, and allow the phenomena to once again become your primary teacher, the appearances may begin to show new and unexpected sides. For example, the pioneering studies of Portmann (1967) and Schad (1977), as well as the more recent work of Lockley (2007) and Riegner (2008), demonstrate that there are underlying non-adaptive morphological patterns in body form, proportions, and appearance (e.g. bird plumage).

Although there have been – both before and after Darwin – biologists and evolutionary thinkers who have interpreted evolution from non-adaptationist perspectives (see Holdrege 2009, for a partial bibliography), these ideas have remained decidedly on the margin and have been largely ignored by the 'normal science' of mainstream biology and biology education. What biology student today would be encouraged to come up with a different theory of evolution than the one that forms the fabric of all her textbooks and courses? Generally speaking, the 'facts' of science are taught without the teacher or professor making the conceptual framework explicit or providing students with the opportunity to look at the same facts from other points of view. The student grows up in the cloak of a theoretical framework and is never taught to take a look at that cloak or given occasion to remove it and try on a different one. When a way of viewing the phenomena becomes the unreflected basis of all investigation then, while certainly leading to knowledge within that limited framework, it limits insight in a broader sense. It limits by cultivating only one kind of approach, and it limits by marginalising other perspectives.

Darwin's theory of evolution is an example of a powerful and widely acknowledged theory. It has stimulated vast amounts of research. At the same time it has decidedly limited the scope of scientific inquiry through its dominance in the scientific establishment. Inasmuch as scientists attempt to apply a theory to every phenomenon it becomes more and more general and abstract. As Ken Wilber remarks in relation to systems theory, 'precisely in its claim and desire to cover *all* systems [it] necessarily covers the least common denominator, and thus nothing gets into systems theory that, to borrow a line from Swift, does not also cover the weakest noodle' (Wilber 2000, p.122). Therefore, he concludes, a general theory may be 'fundamental', but it is also the '*least* interesting, least significant' (p.122; his emphasis), since it only takes into account what fits within the theory. When I state that the scarlet red carapace of a milkweed beetle, the deep blue plumage of a bluebird, and the white fur of an arctic fox are all related to survival, I am in a sense saying something fundamental, but I am also saying nothing specific about those particular animals. When characteristics become no more and no less than survival strategies, then they have become little more than shades of gray in the limited colour spectrum of the general theory. But that is not their fuller reality.

Evolving knowledge

A general theory is essentially something to be overcome as human understanding and science evolve. Because a theory is always a limited human notion, it cannot do justice to the complex nature of the world. The more tentatively guiding concepts are held, the more dynamic and exploratory the mode of inquiry becomes. In dialogue with the phenomenon a new quality of knowing can emerge – the perceptual-conceptual beholding of essential relationships and patterns. These are not abstract ideas and they are also not fixed. They evolve with further inquiry as the human being evolves. When this happens, one can rightfully speak of an organic way of knowing. Knowledge grows out of the careful interaction of human being and phenomenon in away that resembles how the plant develops out of the interaction with its environment. Flexibility of mind, openness to the new, and the ability to let each new phenomenon stimulate the growth of fresh conceptions are the living qualities that characterise an evolving science. This dynamic and transformational nature of inquiry is expressed beautifully by Goethe:

> When in the exercise of his powers of observation the human being undertakes to confront the world of nature, he will at first experience a tremendous compulsion to bring what he finds there under his control. Before long, however, these objects will thrust themselves upon him with such force that he, in turn, must feel the obligation to acknowledge their power and pay homage to their effects. When this mutual interaction becomes evident he will make a discovery which, in a double sense, is limitless; among the objects he will find many different forms of existence and modes of change, a variety of relationships livingly interwoven; in himself, on the other hand, a potential for infinite growth through constant adaptation of his sensibilities and judgment to new ways of acquiring knowledge and responding with action.
> (1995: p.61)

It is precisely this approach to the scientific study of nature that is now so desperately needed if science is to address the disconnect between humanity and the rest of nature that is the root cause of the global environmental crisis.

References

Barfield, O. (1988) *Saving the Appearances,* Middletown, CT: Wesleyan University Press.

Bentz, V., & Shapiro, J. (1998) *Mindful Inquiry in Social Research,* Thousand Oaks, CA: Sage Publications.

Bortoft, H. (1996) *The Wholeness of Nature.* Great Barrington, MA: Lindisfarne Press; Floris Books, Edinburgh.

Cassirer, E. (1971) *Idee und Gestalt,* Darmstadt: Wissenschaftliche Buchgesellschaft.

Darwin, C. (1872) *Origin of Species,* sixth edition. London: John Murray. Downloaded on January 6, 2010, from http://darwin-online.org.uk/contents.html.

—, (1979) *Origin of Species,* reprint of first edition. New York: Penguin Books. (The first edition of this book was originally published in 1859.)

—, (2005) *The Autobiography of Charles Darwin, 1809–1882: with original omissions restored* (ed. N. Barlow), New York: W.W. Norton & Company.

Desmond, A., & Moore, J. (1994) *Darwin: the life of a tormented evolutionist.* New York: W. W. Norton & Company.

Gadamer, H.-G. (1998) *Praise of Theory.* New Haven: Yale University Press.

Goethe, J.W. (1973) *Goethes Werke, Band XII* [*Goethe's works, volume 12*]. Munich: Verlag C. H. Beck. (The translations in the text from this volume are by Craig Holdrege.)

—, (1982) *Italian Journey.* San Francisco: North Point Press.

—, (1995) *The Scientific Studies* (D. Miller, ed. & trans.), Princeton: Princeton University Press.

—, (2010) 'The experiment as mediator of object and subject.' *In Context, 24,* 19-23. (Also available online at: http://www.natureinstitute.org/pub/ic/ic24/ic24_goethe.pdf)

Gould, S.J., & Lewontin, R.C. (1979) , 'The spandrels of San Marco and the Panglossian paradigm: a critique of the adaptationist programme.' *Proc. R. Soc. London B, 205,* 581-598.

Hempel, C.G. (1966) *Philosophy of natural science.* Englewood Cliffs, NJ: Prentice-Hall, Inc.

Holdrege, C. (2005) *The giraffe's long neck: from evolutionary fable to whole organism.* Ghent, NY: The Nature Institute.

—, (2009) 'Evolution evolving.' *In Context,* 21, 16–23.

Kuhn, T.S. (1996) *The Structure of Scientific Revolutions* (third ed.) Chicago: University of Chicago Press.

Lockley, M.G. (2007) 'The morphodynamics of dinosaurs, other archosaurs, and their trackways: holistic insights into relationships between feet, limbs, and the whole body.' *SEPM Special Publication* no. 88 (Society for Sedimentary Geology), pp.27–51.

Maier, G., Brady, R., & Edelglass, S. (2008) *Being On Earth: Practice in Tending to the Appearances.* Berlin: Logos Verlag.

Malthus, T. (1999) *Essay On The Principle Of Population.* Oxford: Oxford University Press.

Portmann, A. (1967) *Animal Forms and Patterns.* New York: Schocken Books.

Ribe, N., & Steinle, F. (2002) 'Exploratory experimentation: Goethe, Land, and color theory.' *Physics Today* (July) Published online: http://www.physicstoday.com/pt/vol-55/iss-7/p43.htm

Richards, R. (2002) *The Romantic Conception of Life: Science and Philosophy in the Age of Goethe.* Chicago: University of Chicago Press.

Riegner, M. (2008) 'Parallel evolution of plumage pattern and coloration in birds: implications for defining avian morphospace.' *The Condor,* 110, 599–614.

Safranski, R. (2009) *Goethe und Schiller: Geschichte einer Freundschaft* [*Goethe and Schiller: the history of a friendship*]. Munich: Carl Hanser Verlag.

Schad, W. (1977) *Man and Mammals: Toward a Biology of Form.* Garden City, New York: Waldorf Press.

Steiner, R. (1996b) *The Foundations of Human Experience.* Great Barrington, MA: Anthroposophic Press.

Tantillo, A.O. (2002) *The Will to Create: Goethe's Philosophy of Nature.* Pittsburgh: University of Pittsburgh Press.

Trout, J.D. (2002) 'Scientific explanation and the sense of understanding.' *Philosophy of Science, 69,* 212-233.

Wilber, K. (2000) *Sex, Ecology, Spirituality* (Collected Works, vol. 6) Boston: Shambhala Publications.

Wolpert, L. (1994) *The Unnatural Nature of Science,* Cambridge, MA: Harvard University Press.

Zajonc, A. (1993) *Catching the Light: the entwined history of light and mind,* New York: Bantam Books.

—, (1999) 'Goethe and the phenomenological investigation of consciousness.' In S.R. Hameroff, A.W. Kasniak, & D.J. Chalmers (eds.), *Toward a Science of Consciousness III* (pp.417–27) Cambridge, MA: The MIT Press.

Selected bibliography of Goethean studies of nature

This bibliography is by no means exhaustive; it is restricted to English language publications and focuses on studies within the biological sciences.

Bockemühl, J., & Suchantke, A. (1995) *The Metamorphosis Of Plants.* Cape Town: Novalis Press.

Colquhoun, M., & Ewald, A. (1996) *New Eyes for Plants.* Stroud, UK: Hawthorn Press.

Goethe, J.W. (1995) *The Scientific Studies.* Ed. and trans. D. Miller. Princeton: Princeton University Press. (This volume includes a variety of Goethe's own studies, including his *Metamorphosis of Plants* and sections of his *Theory of Colours.)*

Hoffmann, N. (2007) *Goethe's Science of Living Form.* Ghent, NY: Adonis Press.

Holdrege, C. (1998) 'Seeing the animal whole: the example of horse and lion.' In D. Seamon & A. Zajonc (eds.), *Goethe's Way of Science* (pp. 213–32) Albany, NY: State University of New York Press.

—, (2000b) 'Skunk cabbage (*Symplocarpus foetidus*)'. *In Context,* 4, 12–18. Available online: http://natureinstitute.org/pub/ic/ic4/skunkcabbage.htm.

—, (2003) *The Flexible Giant: Seeing the Elephant Whole.* Ghent, NY: The Nature Institute.

—, (2005) *The Giraffe's Long Neck: From Evolutionary Fable To Whole Organism.* Ghent, NY: The Nature Institute.

—, (2008b) 'What does it mean to be a sloth?' In C. Holdrege & S. Talbott, *Beyond Biotechnology: The Barren Promise Of Genetic Engineering* (pp.132–53) Lexington: The University Press of Kentucky. Available online: http://www.natureinstitute.org/nature/sloth.htm

Kranich, E.M. (1999) *Thinking Beyond Darwin: The Idea Of The Type As A Key To Vertebrate Evolution.* Great Barrington, MA: Lindisfarne Books.

Riegner, M. (1993) 'Toward a holistic understanding of place: reading a landscape through its flora and fauna.' In D. Seamon (ed.), *Dwelling, Seeing and Designing: Toward a Phenomenological Ecology* (pp.181–215) Albany, NY: State University of New York Press.

—, (1998) 'Horns, hooves, spots, and stripes: form and pattern in mammals.' In Seamon, D., and A. Zajonc (eds), *Goethe's Way Of Science* (pp.177–212) Albany, NY: State University of New York Press.

Schad, W. (1977) *Man And Mammals: Toward a Biology of Form.* Garden City New York: Waldorf Press.

Suchantke, A. (2009) *Metamorphosis: Evolution In Action.* Ghent, NY: Adonis Press.

8. Enlightened Agriculture

COLIN TUDGE

Brian Goodwin, as much as anyone I have known, hit the nails that really need hitting, firmly on the head. He was of course an excellent scientist – one of the rare and valuable breed who brought the special insights of maths and physics to bear upon biology; which, beyond doubt, it needs. But although he loved all of science, both the way it works and the wonders it reveals to us, he also appreciated, as many alas do not, its limitations.

Science, as Sir Peter Medawar put the matter in the mid-twentieth century, is 'the art of the soluble' – no more, no less. Science can do only what science can do. It seems to give firm answers but they are not so firm as they look – that was the greatest lesson of the twentieth century; and they are as firm as they are only because scientists take such care to tailor the questions, in line with their hypotheses and within the bounds of their investigative powers. Scientists cannot address what it does not occur to them to ask – and we can't think of everything there is to think about unless we are already omniscient, which is a logical nonsense.

In practice, too, there are many questions that in principle scientists ought to be able to address, but for technical reasons are unable to – including, in the end, those that relate to the origins of the universe. More broadly: although science can in principle throw at least some light on everything we might want to think about, it cannot tell us what is *right*. In my day, at my university, philosophy was called 'moral sciences' but it truth it cannot be a science in the modern sense. Above all, scientists can never answer the biggest of all questions – '*How come*?' That is metaphysics. Metaphysics can be precise and it certainly makes progress but it clearly isn't science. It does however demonstrate that science does not have a monopoly on truth.

Brian Goodwin took all this for granted. To him it was simply obvious that science is a vital source of insight, yet cannot truly contribute to human wisdom except in constant dialogue with all the other ways by which human beings prehend the universe. We need a constant flow of notions between biology and economics, science and metaphysics, ratiocination and intuition. The way he taught at Schumacher was, I reckon, a model of what education ought to be – truly a route to wisdom; as opposed to training, which is intended only to instil some specific skill or expertise (which in turn, in the present economy, if you are very lucky, might lead to a job with HSBC or Monsanto).

Science should be the great alternative voice – not the royal road to truth to be sure but a necessary and hugely exciting road nonetheless. Today, though, it has become the handmaiden of big industry and big governments. It has, in truth, become seriously corrupted. It needs to be rescued both for its own sake and for the sake of humanity and the world. The rescue requires a different approach to administration and governance – we need more publicly funded science, within a true democracy. But it also requires a re-conception. We need to see that science isn't just a way of making some people more comfortable, and a few people very rich. It really can help us to see how wonderful the universe really is, and help us to live within it, not as conquerors but in harmony. But for this, it needs to be embedded in a far broader conceptual framework, and in particular within a framework of metaphysics.

In short, Brian perceived both the brilliance of science and the limitations of science, and the need for a broad perspective. The plight of modern agriculture and the disasters that have ensued show us what happens when this perspective goes missing.

Shouldn't farming be about feeding people?

I have been looking at agriculture all over the world for the past forty years and along the way I've met many fine scientists – including many who are truly wise. But wisdom, these days, does not prevail in agriculture, and certainly not therefore in agricultural science. Agriculture has been plugged in to a particularly hard-nosed and one-dimensional economy that makes a virtue of its own amorality – what is considered 'right' is what emerges from the market. The kind of agricultural science that gets reported on BBC news is designed and intended not to look after all humanity and the world at large but to maximise the short-term profits of big companies, and to increase the influence of super-powers.

The purpose of agriculture, you might reasonably suppose, is to feed people; and we should try to do that (and in the end we have to do that) without injustice to our fellow human beings, without cruelty to the livestock who are our fellow creatures, and without wrecking the rest of the world. Farming that is designed to do all this has been called 'Enlightened Agriculture': 'enlightened' being a high-fallutin' term for what ought to be common sense and common morality.

The global population now stands at around 6.8 billion and the UN tells us that by 2050 it will reach about 9.3 billion – but then if we follow the demographic curve, numbers should level out. After that the population should start to fall again – not through disaster or war or epidemic but simply because people who are not panicked, and are left to make their own decisions, and have access to contraception, almost invariably choose to have fewer children; and if each couple has only two children then total numbers eventually fall because some of those children for a variety of reasons will have no children of their own.

This – the end of the exponential increase of the human species – is the best news that planet Earth has had for the past ten thousand years, since we started farming in earnest and our numbers started to grow. This also means that for the first time since records began the task of feeding the human species can be seen to be finite. If we can feed 9.3 billion, and go on doing this for a few decades or centuries (after which numbers should fall), and can do this without wrecking everything else, then we've cracked it. Barring asteroids and mega-volcanoes and their like, the human species could look forward to the next million years – and then draw breath and prepare for another million.

If we did set out to instate Enlightened Agriculture – farming designed expressly to feed people without wrecking the rest of the world – then we could surely achieve this. The world is big enough, and the climate and ecological fabric are not yet wrecked beyond recovery (or at least that is a reasonable hope). It should indeed be possible – well within our grasp – to feed everyone who is ever likely to be born on to this Earth to the highest standards both of nutrition and of gastronomy. We should be able to do this with minimum collateral damage – and indeed, as the tide begins to turn again, our farming and our ways of managing land in general should increase the numbers and variety of our fellow creatures, and their long-term security. There isn't space to prove the point here, but we can demonstrate all this by simple arithmetic. A few years ago I wrote a book called *Feeding People is Easy* and I called it that partly because it seemed catchy but mainly because it is true. The amount of food that humanity needs is far outweighed by the amount that reasonably available farmland could reasonably produce. 'Easy' is stretching the point a bit, but not much.

To achieve this, we need very good science – but not the kind of science we have now. The kind we have now, much bruited though it is – GMOs and all the rest – is seriously crude. We need something altogether more subtle. We need, in truth, to move into a new age of biology.

Let us first define the problem. If we are to feed a maximum of 9.3 billion people well, for the next few centuries, and then feed a steadily diminishing but still large population for a few million years after that, we need agriculture that is highly productive – and is also sustainable. But then we know too that conditions will change in the future – which would be true even without the global warming that is already threatening to turn the world upside down. So we need our farming also to be very flexible, or 'resilient'.

Productive, sustainable, and resilient – it seems a tall order. Yet there is a wonderful precedent, all around us. Nature itself. Nature has been wonderfully productive for the past 3.8 billion years – although in that time the Earth has flipped from pole-to-pole ice to pole-to-pole tropics (almost) several times. So how has nature pulled this off? By three outstanding tricks, is the answer. First, it is extraordinarily diverse. Because there are so many species, and often so much genetic variation within each species, lineages of creatures and entire ecosystems can readily change course. That is flexibility.

Secondly, nature is wonderfully integrated. Within populations of particular species, and in ecosystems containing many species, there are many millions – actually it's billions upon billions – of synergies. Charles Darwin emphasised the competition between individuals, and later biologists emphasised inter-species rivalry. Much more striking though, when you look at nature as a whole, is its cooperativeness. Everything benefits from everything else. Thus these maximally diverse populations are also maximally productive. The sheer abundance of pristine nature (if any can be found these days) can beggar belief.

Thirdly, although nature is often said to be prodigal – all that death! – in truth it is a model of thrift. Inputs are minimal, and everything is re-cycled. This is why it has proved so sustainable. Most ecosystems get most of what they need from local sources: the plants set the ball rolling by fixing carbon from the air; the nitrogen comes from the soil, mostly supplied by courtesy of nitrogen-fixing bacteria. Almost all ecosystems also benefit in some way from other ecosystems elsewhere – the oxygen that we breathe on land emanates in large part from photosynthesis in the sea; and the forests of western North America get much of their phosphorus from salmon, swimming upstream from the Pacific and then scattered far and wide by bears and wolves. But everything circulates. Nothing is lost forever (until it finally disappears into the magma, in the fissures between tectonic plates). Only human beings, it seems, scatter non-renewables irretrievably (as we are doing spectacularly with phosphorus).

But farming is not nature. Farming is an artifice. So can these natural principles – maximum diversity, integration, and minimum input – really be translated into farming practice? Of course they can. Diversity, in farming terms, becomes polyculture, usually known as 'mixed farming': many different species of crops and livestock, all of them genetically diverse. The sheer diversity is nature's best defence against pests and diseases. If populations of animals or plants are diverse, then parasites cannot get a foothold. They attack some, but are rebuffed by others. If the parasite does take hold, and causes epidemic, then it will rarely or never kill the whole population. Some among the host population will always be resistant, to carry on the race. Potato blight wiped out the potatoes of Ireland (and western Scotland) in the 1840s because the potatoes were a clone, genetically uniform. Farmers and growers, and particularly organic farmers, who take good care to mix their crops are forever reporting that their crops have more or less escaped the diseases that have laid their industrial neighbours low.

'Integrated' means that the different animals and crops on the farm don't simply live side by side as if in a zoo or botanic garden, but interact. In traditional farms, pigs were the arch-interactors. They fed on swill and even on ordure and hence were waste-disposers. They were often kept on dairy farms to mop up the surplus whey. They cleared the fields after harvest and so were outstanding cultivators. At the end of it all they provided rich manure. Traditionally, indeed, pigs were not kept primarily for food, but for their all-purpose usefulness. The meat was a bonus.

More generally, at least since classical times, there has been obvious synergy between arable crops (notably cereals, formally cultivated on the field scale) and pastoral (grazing cattle and sheep). Traditionally, the arable was in the centre, and the cattle and sheep grazed in the periphery

– their dung ferried in for manure. With the fourteenth century came formal rotations – with livestock and arable taking over from each other year by year. That was true integration. Each crop and class of livestock was enormously various genetically. In recent years the variations in medieval wheat have been strikingly revealed by John Letts, an archaeobiologist turned thatcher. He found a huge mixture of genotypes and phenotypes among the wheat straw of fifteenth century thatches. Now he is growing them again, and making bread from their flour; bringing them back to life (and showing that home-grown wheat, even medieval wheat, makes very good bread, if you bake it properly. It's not suitable for Mother's Pride, but Mother's Pride isn't all there is).

The thrift of nature is mimicked by organic farming. As conceived by Eve Balfour, founder of the Soil Association, organic farming is primarily an exercise in maintaining a good soil structure, rich in organic material – including the ecosystem of soil microbes, fungi, worms, and the rest. But in practice it is also about minimum inputs, and obsessive re-cycling of on-farm materials – the fields are fertilised by clovers and manures, and not with nitrogen 'out of the bag'; the pests are controlled by diversity and rotations, rather than pesticide. In short, a well-managed organic farm is a managed ecosystem.

Farms that mimic nature – mixed, integrated, and low input – are bound to be complex. Minute attention to detail is vital. So they must be labour-intensive. When farms are intricate, minimum input, and labour-intensive, there is little or no advantage in scale-up, and many disadvantages. So the farms that practise Enlightened Agriculture (designed to feed people well forever come what may) are necessarily small. Small farms are best suited to local distribution. That certainly does not mean that every tiny area or even every country should contrive to be entirely self-sufficient, supplying all its own needs from the immediate surroundings. It does mean that all areas should take farming seriously, and produce as much as possible locally.

In structure, such small, mixed, labour-intensive, locally-focused farms are traditional: how most farms used to be and in much of Asia and Eastern Europe, still are. But they are not recommended for reasons of nostalgia. They are what the world needs if we truly want to feed ourselves well, and to go on doing so. In any case, we shouldn't be deceived. The structure is traditional, the farms and their communities are human-sized, and much of the technology may well be traditional too. But the enlightened farms of the future would not confine themselves to old-fashioned knowledge and traditional techniques. Excellent science is needed to make these small, polycultural farms work to their full potential; and small farmers need access to the whole range of modern technologies, tailored to their needs. Overall, however, the science of polyculture has been horribly neglected, and agricultural technology these days is geared to the mega-scale. It all has to be re-created almost from scratch.

If all countries practised this kind of intricate, locally-focused agriculture as a matter of course, then almost all could easily be self-reliant – not producing absolutely everything that people might conceivably want (which is self-sufficiency) but producing enough of the right things to get by on; and this is the essence of food security. Simple arithmetic shows that Britain could easily be self-reliant (as outlined on our website as shown below). The trade in food, whether between districts or between countries or continents, should then become an add-on – a way of cashing in on things a country grows well, and acquiring things that it can't grow at all without huge inputs of energy. All trade should of course be just – 'fair trade' – and should generally be confined to commodities that travel well, and are costly relative to their bulk. Cinnamon and cardamoms certainly qualify, and so do coffee and tea, and fair trade bananas squeeze under the bar. French beans air-freighted from Kenya to Harmondsworth certainly do not. It's hard to improve on common sense.

The notion that almost all countries could be self-reliant if only they designed their agriculture as if they intended to feed their population (without wrecking the rest) may seem surprising. It's the opposite of what we are commonly told these days. But it's true. Notably, most of the countries in Africa that of late have been written off as basket cases could easily provide enough food for their populations, and good food at that, if only their farming was designed for that purpose.

Here we meet a series of serendipities. Farming that mimics nature produces plants and animals in the ratio that nature does, with lots of plants, and not much meat. It is also maximally diverse. This short list – 'plenty of plants, not much meat, and maximum diversity' – summarises all the best nutritional theory of the past thirty-five years. It also summarises the basic structure of all the world's greatest cuisines – Turkish, Indian, Chinese, Italian, Provençale, Lebanese – and so on and so on. In short, sound husbandry, guided by the principles of basic biology, also provides us with the best possible nutrition and is the basis of the finest gastronomy. We are constantly told that to live sustainably we must all tighten our belts. In truth, we merely need to take food seriously (which means re-learning how to cook) and re-awaken the world's traditional cuisines. The future belongs to the gourmet.

In short, Enlightened Agriculture in the end is common sense and common morality. It is rooted in reality – in basic biology. It could and surely would, if practised, produce good food for everyone forever, and take care of our fellow species too. It would solve forever the constant horror of human misery and mass extinction. A huge amount of research is needed to make it work as well as it must – and in this we have to make up for at least half a century of neglect. But technically, feeding all the world's people really should be easy, if that is what we truly set out to do. So why don't we?

Because, of course, the world's agricultural strategy is not rooted in the principles of biology and of common morality. Throughout history, agriculture like every other human activity has been obliged to fit in with the political and economic norms of its day. To be sure, various regimes historically have been sympathetic to farming, and aware of its needs, and then they have commonly left the peasants or the tenants to get on with it. Often, the result came close to realising the ideals of Enlightened Agriculture, as generations of farmers took mixed farming for granted and strove to pass on their land to their children in better heart than they inherited it.

Often, though, especially in the past few centuries, various, bullish political regimes have taken it upon themselves to impose their will, and overridden the basic principles of sound husbandry with ideological dogma. Thus in the 1930s Stalin tried to turn Russia's peasant farms into collective factories to prove that collectivism worked. The prevailing dogma now is that of the neoliberally, notionally free but in truth highly contrived, global market. Stalin deployed the nonsense of Lysenko. The moderns deploy heavy engineering and industrial, basically military chemistry, now abetted by 'biotechnology', in which 'genetic engineering' leading to 'genetically modified organisms' (GMOs) is the jewel in the crown. Both believed or believe that their science made them omniscient, or soon would; and that the high technologies that arose from that science would make them omnipotent, able to manipulate the natural world at will. Both spoke or speak of 'the conquest of nature'. In both cases, the results have been disastrous.

The present, neoliberal economy, the global free market, is maximally competitive (at least in so far as this benefits the big players, otherwise the rules are changed); and is maximally monetised – which means that excellence is measured only in money. All who are engaged in the neoliberal market (and it is very hard indeed to escape) are obliged to maximise profit. They must contrive to make as much money as possible in the shortest time – and if they don't, they will be pushed aside by somebody who does. It all sounds too crude to be taken seriously but this is precisely how it is. To operate according to the diktats of the market is deemed to be 'realistic'. Ideas that merely have to do with feeding people and avoiding extinction by operating within the laws of physics and the principles of biology are deemed to be 'unrealistic'. The abstractions of money are deemed to be more real than the real realities of the physical world.

The demands of the neoliberal market are absolutely at odds with the tenets of Enlightened Agriculture. To maximise profit – in any business, not just in agriculture – suppliers must follow three golden rules: maximise turnover; add value; and minimise costs. This may or may not be a good idea in some contexts. In the context of farming, it is disastrous.

In agriculture, 'maximising turnover' means maximising yield. The modern myth has it that famines occur because traditional crops grown by traditional farmers cannot yield enough. In reality, as the Nobel prize-winning economist Amartya Sen and others have pointed out, this is almost never the case. Famines are almost always caused by extraneous political forces, such as civil war, or by economic policies that separate the people from their own land. Agriculturalists who understand Third World farming, such as Professor Bob Orskov of the Macaulay Institute, Aberdeen, say that almost all traditional farmers could produce far more than they tend to do, using their own methods, if only conditions were right. For instance, dairy farmers in Indian villages could often produce enough milk to sell to the local towns – but they don't because in the rainy season, when the cows are at their milkiest, they can't get to market because the roads are too muddy. So they need better roads or better carts. What they emphatically don't need is what western scientists commonly recommend or even foist upon them – 10,000 litre Friesian cattle bred to feed on custom-bred ryegrass and imported soya. Worldwide, frantic attempts to maximise cereal yields in the short term with more and more fertiliser and deeper and deeper ploughing commonly destroy soil structure and create desert or swamp. But that is what the market demands.

Value adding means packaging and freezing and sending out-of-season fruit across the world, which is obviously gross. It also means feeding grain and other staples that could be the basis of great cuisine, to livestock that should be living on grass or leftovers: half the world's wheat, 80 per cent of the maize, 90 per cent plus of the soya (grown mostly in Brazil at the expense of the rain forest and the Cerrado). Livestock raised by traditional means enhance the efficiency of farming, by feeding as sheep and cattle traditionally do in places where we can't grow arable crops; or thrive on surpluses and leftovers that we don't want to eat, which is the traditional role of pigs and poultry. Grain-fed livestock compete with humanity for food. By 2050 on present trends cattle and pigs will be gobbling enough staple to feed another four billion people – increasing the effective world population to thirteen billion. No wonder the powers-that-be are panicking.

Even worse, is the perceived need to reduce costs. Britain's livestock have suffered almost continuous epidemics since the 1970s – BSE, the worst ever outbreak of foot-and-mouth disease, swine fever and swine flu, the narrowly averted (so far) threat of bird flu. Meanwhile, the perpetual leitmotiv, TB rumbles on, prompting spasmodic government attacks on badgers. Almost all of this horror can be ascribed to cut-price husbandry: the lack of surveillance; the far-flung abattoirs, with animals ferried hundreds of miles; the leaky national borders; the cheap imports; and piecemeal stabs at control that in large part are window-dressing.

Worst of all, though, is the perceived need to reduce labour – the most costly input in traditional systems. In 'advanced' countries, agriculture is 'industrial': farm labour is replaced by big machinery and industrial chemistry. In Britain and the US, only one per cent of the work-force is full-time on the land. In the world as a whole it is nearly 50 per cent. Britain and the US are seeking to impose their own methods on the world at large in the interests of 'modernity'. If they succeed, half the world will be out of work. Already a billion people live in urban slums and many and probably most of them are dispossessed farmers and their immediate families.

With the work-force reduced, and reduced again, complex husbandry becomes impossible. Hence today's farming is as simple as possible. This is why we have monocultures as far as the eye can see; and livestock raised not on the fields and in the farmyards but in factories – dairy factories with many thousands of animals; pig factories with up to a million animals; poultry factories with several million; all raised on grain and soya. The manure that should be a major asset to fertilise the fields becomes a huge embarrassment. Biologically, the manure from a million-head piggery is equivalent to the sewage output of London.

All this is the proximal cause of disaster. It is hugely profligate and hence unsustainable (it all depends on oil, and non-replaceable resources such as phosphorus are scattered to the four winds); and hugely damaging (destroying soils, polluting oceans, squandering fresh water, and a major cause of global warming); a frantic exercise in short-termism.

The economy is the driver. But beneath the economic nonsense is a deeper cause: the misreading both of nature (how it really works) and of science (what it can really tell us and what it can't). For although industrial farming looks so modern in the eyes of the powers-that-be, in truth it is primitive. The belief that we can indeed achieve omniscience, and understand exhaustively how living nature works, belongs to the eighteenth century, when recognisably modern science was still new. The idea that we can 'conquer' nature, mould it to our will, belongs to the nineteenth century, when heroic engineers like Isambard Kingdom Brunel used enormous quantities of iron and steel to create structures that seemed to suggest that human ingenuity can indeed achieve anything.

For the most part the twentieth century continued on this heroic path – which is how we landed up with industrial agriculture. But twentieth century also showed us the limitations of science: how far we must always fall short of omniscience, and hence of omnipotence. It also revealed the limits of the world's resources and the inescapable need for thrift.

Above all, perhaps – and this is the lesson that came late in the century, from physics – it showed us that nature as a whole does not follow simple cause-and-effect rules, in the manner of Newton's mechanics. Cause and effect in nature above all is 'non-linear'. We may set things in train but we cannot determine or precisely predict the course of events. Nature cannot simply be tailored to fit. It cannot be 'engineered'. We cannot simply apply the rules of the production line. At best, nature can be orchestrated. This is not fanciful romanticism. This is real science – the kind of science we need in the future. The science and high-tech that are now supposed to be so modern, the agrochemistry and the biotech, in truth belong conceptually to times long past.

Brian Goodwin knew all this. Non-linearity and all that went with it was his area – complexity; self-organising systems. He also pondered the relationship between biology and economics (what the latter can learn from the former). Above all, he had that intuitive feel for nature, that empathy for it and respect for it, that has much more to do with metaphysics than with science but provides the essential grounding for science nonetheless. Nature may be orchestrated, up to a point, but it cannot be controlled. The dream of conquest is both vile and ludicrous. The goal is harmony. All this, Brian took to be self-evident. The world needs his way of thinking, and the kind of education, broad and open-ended, that he took part in.

References

Goodwin, Brian (2007) *Nature's Due,* Floris Books, Edinburgh.

Tudge, Colin (2008) *Economic Renaissance,* Schumacher College paper No.1, Schumacher College, Dartington, UK.

9. Two Practical Suggestions for Improving Scientific and Medical Research

RUPERT SHELDRAKE

The way science is currently organised and taught serves to reinforce the mechanistic worldview. And since this view of nature as inanimate and purposeless is a major ingredient in the ecological crisis, science is part of the problem.

For a start, we need a more holistic approach to science education both in schools and universities, and that is an area where Schumacher College is already playing an important role, and to which Brian Goodwin made such an important contribution.

We also need a major paradigm shift within science itself towards a new vision of living nature, and again Schumacher College has been playing an important part in this process, with Stephan Harding's book *Animate Earth* (2009) a significant milestone.

But within the vast majority of educational establishments and research institutes, it is business as usual, and what happens at Schumacher College has little immediate effect, although hopefully it is making a long-term contribution to change.

I propose here two suggestions for the reform of research. Both are modest in scale, and quite practicable, and would, I believe, have far-reaching effects on the practice of science and medicine.

The first concerns science funding, and arose out of a conversation I had on this subject with Satish Kumar at Schumacher College. The second proposes a simple reform of medical research that would make alternative and complementary therapies easier to test. The better established such therapies become, the more they will highlight the need for a holistic approach in the life sciences.

Democratising science

Science has always been élitist and undemocratic, whether in monarchies, communist states or liberal democracies. But it is currently becoming more hierarchical, not less so, and this trend needs remedying.

In the nineteenth century, Charles Darwin was just one of many independent researchers who, not reliant on grants or constrained by the conservative pressures of anonymous peer review, did stunningly original work. That kind of freedom and independence has become almost non-existent. These days, the kinds of research that can happen are determined by science funding committees, not the human imagination. What is more, the power in those committees is increasingly concentrated in the hands of politically adept older scientists, government officials and representatives of big business. Young graduates on short-term contracts constitute a growing scientific underclass. In the US, the proportion of biomedical grants awarded to investigators under 35 plummeted from 23 per cent in 1980 to 4 per cent today.

This is bad news. As science becomes more and more about climbing corporate career ladders, and less and less about soaring journeys of the mind, so the public's distrust of scientists and their work seems to grow.

In 2000, a government-sponsored survey in Britain on public attitudes to science revealed that most people believed that 'science is driven by business – at the end of the day it's all about money'. Over three-quarters of those surveyed agreed that 'it is important to have some scientists who are not linked to business'. More than two-thirds thought 'scientists should listen more to what ordinary people think'.

Worried about this public alienation, the British government now says it wants to engage the wider public in 'a dialogue between science, policy makers and the public'. In official circles, the fashion has shifted from a 'deficit' model of the public understanding of science – which sees simple factual education as the key – to an 'engagement' model of science and society.

This may be helpful, but it is not enough. There needs to be a different way of funding research. I suggest an experiment: spend one per cent of the science budget on research of real interest to lay people, who pay for all publicly-financed research through taxes. Then science would literally become more popular.

What questions would be of public interest? Why not ask? Organisations such as charities, schools, local authorities, trades unions, environmental groups and gardening associations could be invited to make suggestions. Within each organisation, the very possibility of proposing research would probably trigger off far-ranging discussions, and would lead to a sense of involvement in many sections of the population.

To avoid the 1% fund being taken over by the science establishment, it would need to be administered by a board largely composed of non-scientists, as in many research charities. Funding would be restricted to areas not already covered by the other 99% of the public science budget.

This system could be treated as an experiment, and tried out for, say, five years. If it had no useful effects, it could be discontinued. If it led to productive research, greater public trust in science and increased interest among students, the percentage allocated to this fund could be increased.

The National Center for Complementary and Alternative Medicine (NCCAM), established by the US Congress in 1998, sets a precedent. Complementary and alternative medicine are of great interest to millions of Americans, and NCCAM's current annual budget is $100 million. But before NCCAM's predecessor, the Office of Alternative Medicine, was set up by Congress in 1992, research in these fields was receiving practically no support through established agencies. This is still the case in most countries, including Britain.

This new venture, open to democratic input and public participation, would involve no additional expenditure, but would have a big effect on people's involvement in research and innovation, and help break down the depressing alienation many people feel from science. It would enable scientists themselves to think more freely. And it would be more fun.

Level playing field research in medicine

In medical research, the 'gold standard' research methodology involves randomised double-blind placebo-controlled trials. These trials are helpful in distinguishing the effects of a treatment from the effects of a placebo, but they do not provide the information that is needed by many patients and health care organisations. For example, if I am suffering from lower back pain, I do not want to know whether drug X works better than a placebo in relieving this condition, but which kind of treatment I should seek out of the various available therapies: physiotherapy, acupuncture, osteopathy, and so on.

Probably the best way to answer this question would be a 'level playing field trial' in which various possible treatments were compared with each other. Taking the example of lower back pain, in such a trial a large number of sufferers, say 1,200, would be allocated at random to a range of treatment methods. Five treatment methods could be included in the trial, plus one no treatment group. Thus for each method there would be two hundred patients. The treatment methods could include physiotherapy, osteopathy, acupuncture, chiropraxis, and any other therapeutic method that claimed to be able to treat this condition. Within each treatment group, there would be five different practitioners, so that in the statistical analysis the variability between practitioners could be compared.

The outcomes would be assessed in the same way for all patients at regular intervals after the treatment. The relevant outcome measures would be agreed in advance in consultation with the therapists involved in the trial. The data would then be analysed statistically to find out:

—which treatment, if any, worked best;
—which treatment methods had the greatest inter-practitioner variability;
—which methods were the most cost-effective.

This kind of information would be of great use to patients and also to providers of health care such as the National Health Service.

A similar level playing field approach could be adopted for a variety of other common conditions, including migraine headaches and cold sores.

This would be genuine evidence-based medicine, the trials would be relatively simple and cheap to conduct, and the exercise would be pragmatic and theory-free.

Imagine, for example, that homeopathy turned out to be the best treatment for cold sores. Some might argue that this was simply because homeopathy brought about a stronger placebo effect than the other treatments. But if homeopathy unleashed a greater placebo effect than other methods, then this would be an advantage, not a disadvantage.

Outcome research of this kind used to be common in medicine before the Second World War, and it is still widely used as a research in other areas, for example in agricultural field trials, and also increasingly in research on the effectiveness of different kinds of psychotherapy. Standard statistical methods can be used in the analysis of data.

Level playing field outcome research comparing different treatment methods, including complementary and alternative therapies, would be helpful both for health care providers and for sufferers who are trying to decide which method of treatment to go for. It would also help to bring alternative and complementary therapies within the purview of science, and would hopefully lead to new scientific questions that go beyond the mechanistic framework of most current research.

References

Harding, S.P. (2009) *Animate Earth: Science, Intuition and Gaia.* Green Books, Dartington, UK.

Sheldrake, R. (1994) *Seven Experiments That Could Change the World,* Fourth Estate, London, UK.

10. Ancient Futures

HELENA NORBERG-HODGE

I am convinced that the solutions to our many social and environmental crises in both North and South are more easily attainable than most people believe. However, in order to identify them, we first need to be able to see the 'big picture' – to examine the realities of conventional economic development with open eyes.

Myths about growth and progress, along with romanticised images of urban consumer culture have pervaded nearly every society on the globe; yet these can be counteracted if we let real insight and first-hand knowledge shape our vision of the future. If we really want to achieve the goals espoused by development – namely to relieve poverty and increase human wellbeing – we urgently need to expose two key assumptions that stand in our way:

1. Growth through global trade is necessary to increase employment and reduce poverty.
2. Large-scale industrial agriculture is necessary to feed the world.

From Mongolia to Patagonia, people are led to believe that farming is a thing of the past, that learning English is necessary for their economic survival, that global brands are superior to local ones, and that it is natural for competition, stress and time pressures to keep escalating. There is a tendency to believe that these pressures, along with most of the crises we face, are simply due to a combination of over-population and innate human greed. My time in Ladakh showed me not only the fallacy of these assumptions, but most importantly convinced me that we can find a different and better way forward.

It was in 1975 that I first went to this remote Himalayan region, as part of a film crew making a documentary on the Ladakhi culture.

Politically, Ladakh is part of India, but previous to the 1970s the area had been sealed off from the modern world for strategic reasons. I was working as a linguist at the time and was to learn Ladakhi and act as translator. Through learning the language I gained somewhat of an 'insider' perspective and was warmly welcomed into Ladakhi society. I came to know a people who had not been colonised and were still living according to their own values and principles. Despite a harsh and barren environment of extreme temperatures, the Ladakhis were prospering materially, but also, and even more importantly, emotionally. Over time, I came to realise that they were among the freest, most peaceful and joyous people I had ever met. I also discovered that their happiness translated into a remarkable tolerance, an acceptance of difference and of adversity.

I quickly saw that my western beliefs about a life close to the land involving non-stop toil and drudgery were wrong. In traditional Ladakh, the ratio of people to each parcel of land was such that farm work was accomplished relatively easily, at a leisurely pace. Song, dance and celebration accompanied the work even during the harvest, the peak working season.

The local economy was founded on cooperative small-scale agriculture, which not only provided for people's food needs it also supported an interdependent, close-knit community structure that connected individual to individual and village to village across the region. Ladakh was not a paradise; people, of course, faced problems and hardships. However, I know from personal experience that it was very different from the pictures of hard labour, ignorance, illiteracy, and ill-health propagated by the proponents of development.

Around the time that I arrived, the region was opened up to economic development. Suddenly, the Ladakhis were exposed to outside influences they had not encountered before: advertising and corporate media, tourism, pesticides, western-style schooling, and subsidised consumer goods. The process of development centralised political power in Leh, the capital, and created dependence on an outside money economy for people to meet even their most basic needs. Because of standardised education, tourism and glamourised images in the media, Ladakhis began to think of themselves and their culture as backward and inferior. They were encouraged to embrace a Western urban lifestyle at all costs.

Over the next twenty years I saw the small capital town of Leh turn into an urban sprawl. The streets became choked with traffic, and the air tasted of diesel fumes. 'Housing colonies' of soulless, cement boxes spread into the dusty desert. The once pristine streams became polluted, the water undrinkable. For the first time, there were homeless people. Villages were drained of life as farming became uneconomical and the young went to the cities in search of scarce jobs. Within a few years, unemployment and poverty, pollution and friction between different communities appeared.

The process was so rapid and clear-cut in Ladakh that it was impossible to deny that economic development had created serious social and environmental problems. By the mid-eighties, my organisation and I had been invited to work in other cultures, including Bhutan. A decade later we had contact with dozens of other regions and institutions, because my book and film about development in Ladakh, *Ancient Futures,* had been translated into over forty languages. Again and again we were told, 'the story of Ladakh is our story too' – to varying degrees the same destructive process was at work in almost every other society.

From colonialism to development to globalisation

Today's development approach has its roots in colonialism and slavery. In the early years, European countries spread their influence through robbing other peoples of their land and labour. A European élite forced native peoples to provide for distant markets rather than for their own needs. Whole countries were turned over to the production of single commodities: bananas, coffee, copper. Large-scale agricultural production for export was favoured over diversified, localised farming. In the process, trading corporations expanded and became more and more powerful.

This was followed by an era of so-called 'independence' and 'development'. During this period, former colonies were considered to be in control of their own affairs. However, the economic enslavement continued. A new westernised élite was encouraged to continue providing for export markets rather than for their own people's needs. This enslavement was, and still is, reinforced by ever increasing debt. Today this process continues in fundamentally the same direction – under the banner of economic globalisation.

Most development projects today are well-intentioned, but fail to achieve any kind of measurable progress in the global South. This is not because they lack funding or government attention, but rather because they are embedded in an economic framework that is, quite simply, antithetical to human wellbeing. To illustrate this, the UK gave over £7,000 million in 'Overseas Development Assistance' in 2009 – a record amount of aid.[1] Yet in the same year, the number of people living in extreme poverty increased by some fifty-five million.[2]

Nevertheless there is a widespread belief that globalisation is bringing the world together and necessary to create prosperity for all. The disconnect results in part from the way governments measure 'progress'. For example, we compare the material wealth in the West with what was available fifty or one hundred and fifty years ago. More often, the baseline from which comparisons are made is rooted in the Dickensian period of the early industrial revolution, when exploitation and deprivation, pollution and squalor were rampant.

Similarly, the baseline in the Third World is the immediate post-colonial period, with its uprooted cultures, poverty, over-population and political instability. Based on the misery of these starting points, political leaders can argue that our technologies and our economic system have brought a far better world into being. However, if we step back and take a good look at the 'bigger picture' we gain a very different view.

Global growth does not equal real development

Fundamentally, globalisation is about the deregulation of global trade and finance, doing away with the rules and regulations that protected natural environments and human societies from overexploitation. Bilateral and multilateral 'free' trade treaties allow corporations to move in and out of local economies freely – seeking regions where regulations and costs are lowest.

We are now in the clutches of an economy driven by speculation, consumerism and maximising profits at all costs. The same uncontrolled speculation that led to the recent international financial crisis, where upwards of thirty million people lost their jobs,[3] is also behind the ruthless exploitation of natural resources to the detriment of both human and ecological health. When blind, speculative, money controls what happens in the economy, democracy is undermined and we all lose out. Yet the solution offered by governments worldwide is to encourage us to consume ever more to keep the system afloat.

Globalisation also affects our very sense of wellbeing. Just as in Ladakh, the psychological impacts are apparent around the world – rates of depression, suicide and violence are all on the increase, in North and South alike. In a survey of more than sixty-five countries conducted from 1999 to 2001, Nigeria turned out to have the highest percentage of people who considered themselves happy. Britain ranked twenty-fourth on this scale, despite boasting a GDP more than twenty-two times higher than that of Nigeria.[4]

A few years ago, a BBC poll found that 81% of people surveyed thought the government should focus on making them happier rather than wealthier.[5] However, though some governments are now beginning to look at the issue of human wellbeing, policies are still geared towards promoting globalisation. It's not that policy makers are ill-intentioned; rather that the system itself has blinded them to the real impacts of their actions. Gross Domestic Product (GDP) measures only economic activity – whether beneficial or negative. For instance, the expenses involved in treating illness and pollution, in rebuilding after war ... all of these activities contribute to GDP and are therefore seen as contributing to 'progress'. The numbers investors manipulate on a computer screen in New York City may have their most severe consequences in rural Thailand or sub-Saharan Africa. It is a system that is both blind to its impacts and blind to its own blindness.

Food and farming at the crux of development

Proponents of globalisation argue that unlimited global trade increases efficiency. They consider it more efficient to eradicate local and regional trading systems. More efficient to standardise crops so that everyone eats the same variety of GM corn, the same variety of hybridised rice, the same soft drinks, the same candy bars. More efficient to fly apples from the UK to South Africa to be waxed and washed and then flown back to the UK to be sold in supermarkets. More efficient for Spanish markets to sell Danish butter, while Danish stores sell butter produced in France.

Even foods that appear to be local may have travelled a great distance. For instance, Scottish prawns are shipped to China to be shelled then shipped back to Scotland where they are breaded and sold. Haddock, caught by British trawlers in the Atlantic, goes to Poland for processing, and then back to Britain for sale. Welsh cockles find their way to Holland to be pickled and canned before winding up on UK supermarket shelves.

In an era of dwindling fossil fuels and increasing carbon emissions, this kind of wastefulness is nothing short of madness.

However, these examples of wasteful trade are direct results of the global economy's 'rules of the game', which enable businesses to profit by exploiting wage differences, currency fluctuations, subsidies, and speculative bubbles wherever in the world they are most advantageous.

In the industrialised world, the majority of the population has been herded into a suburban or urban existence. They have lost touch with the land and sources of their food, and as a consequence, have been easily manipulated into believing a big business view of the world.

We have been led to believe that large farms produce more food than small ones. This is an understandable misconception, because on the surface it seems obvious that a larger farm produces more food. However, it turns out that small diversified farms generally produce much more food per unit of land than large industrial monocultures. Numerous studies over the last sixty years have shown that from the US to Malaysia, India to Brazil, Nepal to the Philippines and everywhere in between, there is almost an inverse relationship between farm size and productivity. In Turkey, for example, farms of less than one hectare were found to be twenty times more productive than those of over ten hectares.[6]

The great gain of modern, corporate agriculture has been an 'efficiency' that we no longer can afford: namely that it produces food with less labour. This means that it provides fewer jobs because fossil fuels and machinery have replaced human labour. The end result is a global food system that produces less food, fewer jobs and vastly more pollution.

The global economic model systematically disassembles the structure of the rural economy, leaving many unable to provide for themselves, their families and communities. People are pulled off the land into urban areas where they are forced to compete for scarce, low-paying, insecure employment – away from assured subsistence to dependence on an exploitative globalised economy. When people are pulled into urban centres they lose the connection with others that stems from local bonds of interdependence. Additionally, competition for scarce jobs pits people against one another, creating tensions and even outbreaks of violence. The fact is that destroying rural livelihoods in the so-called 'developing' world is a recipe for famine, conflict and massive increases in pollution. On the other hand, if the bulk of what people ate came from their region, or even from their own country, then millions of farmers and businesses would be making a profit, including those in the South.

Education for development

Education is one of the most powerful tools of globalised development. No one can deny the value of real education, that is, the widening and enrichment of knowledge, but today education has become something quite different. Like colonialism, development standardises the process of education according to a Western model. It isolates children from their culture and from nature, training them instead to become narrow specialists in a westernised urban environment. This was especially clear in Ladakh where I saw that modern schooling acts almost as a blindfold, preventing children from seeing the context in which they live. They leave school unable to use their own resources, unable to function in their own world.

For generation after generation, Ladakhis grew up learning how to provide themselves with clothing and shelter; how to make shoes out of leather and robes from the wool of their animals; how to build houses out of local wood, stone and earth. Children were given an intuitive awareness that allowed them, as they grew older, to use resources in an effective and sustainable way. None of that knowledge is provided in the modern school. Children are trained to become specialists in a technological, rather than an ecological, society. School is a place to forget traditional skills and, worse, to look down on them.

In every corner of the world today, the process called 'education' is based on the same assumptions and the same Eurocentric model. The focus is on faraway facts and figures, a supposed universal knowledge. The books propagate information that is meant to be appropriate for the entire planet. But since only a kind of knowledge that is far removed from specific ecosystems and cultures can be universally applicable, what children learn is essentially synthetic, divorced from the living context. If they go on to higher education, they may learn about building houses, but these houses will be of concrete and steel, the universal box. So too, if they study agriculture, they will learn about industrial farming: chemical fertilisers and pesticides, large machinery and genetically modified seeds. Modern education has brought obvious benefits, like improvements in the rate of literacy and numeracy. However, in the case of Ladakh and elsewhere it has encouraged competition between people, divided them from the land and put them on the lowest rung of the global economic ladder.

Advertising and media further compound the effects of education – at all ages, but especially in the young. The psychological pressures on them are enormous – adverts, satellite television, magazines, billboards and the radio promise them that they will 'belong' and be loved and admired if they wear a certain brand of clothing and have the latest techno-gadget.

What real development would look like

Fortunately, we still have the opportunity to transform development into something that brings us closer to the land and to each other, that provides amply for everyone's material and psychological needs. My own view of what this would look like has been profoundly influenced by ongoing experiences with people living close to the land: farmers, permaculturists and indigenous people. The kind of localised development I advocate would allow us to foster a multi-polar world where the diversity of languages, practices and races, would have a place and an equal voice. Communication would radiate out from each and every one of millions of cultural centres. The use of renewable energy technologies would be tailored and adapted to different ecosystems with varied climates and topography. Our cities would be smaller in size and maintain a relationship with the land around, so there would be a better balance between urban and rural.

There would be ample and meaningful employment – getting more people involved in caring for the earth and for one another. Whether farming, forestry or fishing, medicine, architecture, education, or counselling, there would be no limit to work opportunities to ensure that these activities are carried out with intelligence and care. People would be able to re-connect with the sources of their food, which would have significant economic and health benefits. Perhaps most importantly, we would have stronger, more resilient and closely knit communities, which inspired a sense of belonging and purpose – essential for psychological wellbeing.

If we can release our creative potential and skills from the stranglehold of the corporate consumer culture we can meet our real needs without undermining the life support systems of the planet or homogenising diverse societies.

Realising that it is not human nature or over-population that is the major cause of our crises, but rather an inhuman system, can be immensely empowering and inspiring. As more people become aware of this, we are seeing broad-based support – from social as well as environmental movements – for a fundamental shift in direction. I call that shift economic localisation. Localisation is a far more effective strategy for addressing poverty, social and environmental problems than conventional development and globalisation. All around the world, people are beginning to understand that we need to localise, rather than globalise, our economies.

Localisation for genuine development

In order to create the structures that support genuinely sustainable development we need to rebuild the social fabric, to rebuild community. At a structural level this means restoring economic interdependence at the local level. Human-scale ways of meeting our needs allow us to see our impact on others and on the natural world. We need to adapt economic activity to place, shortening the distance between producers and consumers. Producers can respond to the needs of the consumers, and take responsibility for what they produce as they see the effects immediately before them.

Localisation means greater self-reliance, but it doesn't mean eliminating long distance trade; it simply means striking a healthier balance between trade and local production. It is a process that inherently nurtures a sense of connection to community and to the earth. Strong local economies are essential for helping us to rediscover what it means to belong to a culture, a community, a place on earth.

If we implemented localisation on a global scale, we could actually eliminate hunger and poverty worldwide. Localisation fosters an interdependent network of strong communities and local economies rather than a collection of countries in competition with one another and dependent on a volatile economic system. There will always be changes in local conditions that bring about ecological instability and social problems. However, if economies are adapted to the local environment and the needs of the community, if people's ties to one another are unbroken, if people have free access to their own natural resources, then they are far more likely to recover quickly from any crisis.

Over the last thirty years my organisation has worked with Ladakhi leaders to protect and rekindle these connections. We have collaborated with the youth and farmers, traditional doctors and women's organisations, politicians and business owners to rebuild a sense of pride in the Ladakhi culture and way of life. We have set up appropriate technology projects that use Ladakh's natural resources sustainably, such as small-scale hydro and solar energy, greenhouses and solar ovens. We also run educational campaigns that provide the Ladakhis with a fuller picture of life in West, including the negative sides not shown in the media.

Although these are small steps and the work goes slowly, we are beginning to see the benefits of our efforts. More and more Ladakhis are interested in keeping their spiritual and ecological values alive. Farmers are now more aware of the risks of chemical agriculture and genetically modified crops. There has also been a resurgence of interest in traditional methods of healing. Our most successful collaboration has been with the Ladakh Ecological Development Group and the Women's Alliance of Ladakh, both of which we helped to found. These two organisations are now among the most powerful forces for positive change in the region. They have been promoting renewable energy, organic agriculture, handicrafts and, most importantly, respect for Ladakh's ecological and community values.

The story of tradition, change and renewal in Ladakh has captured the hearts and minds of countless people from Mongolia to the USA, from Burma to my native country of Sweden. As a consequence, I've been in touch with literally thousands of individuals and projects that are reweaving the fabric of local cultures, communities and economies. These movements are rooted in people's desire to preserve the bonds to family, community, and nature that make life meaningful. However, to implement localisation, policy change at the national level is essential as well, and ultimately requires an overall international framework that fosters grassroots change.

All around the world, people are demonstrating incredible wisdom, courage and perseverance in resisting the globalised economic path and formulating more positive approaches to improving quality of life. It is my conviction that if we joined together to localise on a global scale, we could bring about tremendous and lasting positive change. For if we really want to address poverty, hunger, inequality and environmental crises, localisation offers a much greater chance of success than the conventional development path.

Notes

1. DFID (2010) *Statistics on International Development.* Available at: http://www.dfid.gov.uk/Documents/publications1/sid2010/SID-2010-Statistical-Release.pdf.
2. Ellmers, B. (2010) *Official Development Assistance 2009: Poverty on the up as EU aid falls.* Available at: http://www.betteraid.org/en/news/aid-and-europe/265-official-development-assistance-2009-poverty-on-the-up-as-eu-aid-falls.html.
3. Deen, T. (2010) *30 Million Lost Jobs Drag Down Global Economic Recovery.* Available at: http://ipsnews.net/news.asp?idnews=53798
4. BBC (2003) *Nigeria tops happiness survey.* Available at: http://news.bbc.co.uk/1/hi/3157570.stm
5. Easton, M. (2006) *Britain's happiness in decline.* Available at: http://news.bbc.co.uk/1/hi/programmes/happiness_formula/4771908.stm
6. Monbiot, G. (2008) *Small is Bountiful.* Available at: http://www.monbiot.com/2008/06/10/small-is-bountiful/

References

Norberg-Hodge, H. (1992) *Ancient Futures,* Sierra Club Books, San Francisco, CA.

11. Holism and the Reconstitution of Everyday Life: A Framework for Transition to a Sustainable Society

GIDEON KOSSOFF

Introduction: the need for a framework for transition

No era in human history has been so beleaguered by such a range of mutually exacerbating problems: from loss of biodiversity to social injustice; from the proliferation of waste to cultural homogenisation; from rampant urbanisation to nuclear proliferation; from the demise of community to global warming. The list of problems could be extended *ad infinitum,* no aspect of our lives or that of other species remains untouched, and any such problem represents a challenge to the sustainability of human and non-human life support systems. On the other hand, possibilities of a different kind of society and a different way of life are evident in many initiatives that are taking place all over the planet, including renewable energy technologies, local currencies, experiments in participatory democracy, efforts to restore forests and watersheds, new forms of transportation and new kinds of manufacturing facilities, to name just a few. As the list of 'problems' can be extended *ad infinitum,* so too can the list of possible 'solutions'. And whilst we may feel overwhelmed by what historian Russell Jacoby refers to as a state of *'permanent emergencies'* (Jacoby 2005, p.ix), viable (albeit fragile) alternatives to 'business as usual' are not difficult to find.

In this chapter I argue that we need a new framework – a conceptual structure which provides the basis for action in the world – to assist the process of transition to a sustainable society, that is, the process by which we address the kinds of problems touched on above. I further argue that the key elements of such a framework can be realised by applying the insights of whole systems science and philosophical holism to human affairs. Several important 'green' frameworks have already been developed, but these tend to focus on single aspects of the transition process. For example, the framework of Libertarian Municipalism, principally developed by social ecologist Murray Bookchin (Biehl & Bookchin 1998), makes the case for direct and participatory democracy and focuses on the problem of political disenfranchisement and disengagement that is endemic to representative democracy. The Natural Capitalism framework, developed by physicist Amory Lovins and colleagues (Hawken, Lovins & Lovins 1999), focuses on technological issues and advocates an expansion of the definition of 'capital' to embrace ecosystem services and natural resources. The objective is the development of new manufacturing processes that will lay the foundation for the 'Next Industrial Revolution'.

A notable exception to the limited purview of such frameworks, and one of most important developments on the sustainability scene in recent years, is the approach that is being developed by the Transition Town Movement (Hopkins 2008). This is a grassroots effort to enable communities to recover their 'resilience' (Walker & Scott 2006), their capacity to be self-reliant, materially, culturally and psychologically, so that they can flourish not only under ordinary circumstances, but also in the event of external disruptions such as the prospective decline in oil availability. Their systemic and community based approach has the potential to integrate the many different facets of transition – political, technological, cultural, social, economic, and ecological. This is an enormously ambitious project – no less than the transformation of industrial civilisation. Given the enormity of this ambition there are many questions that the Transition Town approach is not equipped to address, such as how to think about the appropriate levels of scale which correspond to the levels of scale for solutions, and the need for a narrative that connects the diverse efforts currently underway all over the planet, including the 'undeveloped world'.

Several additional and essential features of a framework for transition are:

— it needs to incorporate a vision of a future, a desirable sustainable society by which we can orient ourselves in the present;
— it needs to provide a conceptual model for transdisciplinary collaboration (since the expertise required for transition

will come from all fields) within a grassroots context (if the transition movement is not going to become co-opted by experts);
— it needs to provide a way for projects and practices to be connected and integrated, since these will only realise their potential through such mutually beneficial relationships; and
— it needs to embody a more qualitative and humane understanding of sustainability than recent technocratic and economistic appropriations of this concept have come to do.

These seven or eight points lay out an ostensibly overwhelming set of requirements for a framework for transition. The answer, I believe, is to apply a holistic paradigm to how we live – to everyday life. In so doing, a framework which meets all these criteria will emerge. In the following sections of this chapter I will discuss the tradition of holism in social theory and the implications that contemporary whole systems science and philosophic holism have for this tradition, and use this discussion as a basis for a framework that can be integrated into our everyday lives.

Social holism and social sciences

Within social theory, holism (or organicism – I use these words interchangeably) has been one of the most common ways of describing and interpreting human affairs (Scott 1991, Phillips 1976, Stark 1962, Merchant 1983, Hollis 1994, Polkinghorne 1983, pp. 135–67, Harvey Brown, pp.129–39). Holism's influence has been immense even when the originating metaphor – of society as a living and growing organism and therefore an irreducible unity which cannot be understood by reduction to its 'parts' – has long been forgotten. As sociologist Richard Harvey Brown (1940–2003) contended, 'This biological image of society is so deeply rooted that scholars often fail to recognise it as the central presupposition of their own social thought. But, recognised or not, the metaphor can be seen in the substructure of the vast majority of Western theories of social order and change' (Brown 1989, p.131). But it has been a problematic metaphor whose influence has often been reactionary or authoritarian. 'Functionalism', for example, a theory originated by

Herbert Spencer (1820–1903) in the mid-nineteenth century but carried forward by other sociologists and anthropologists for at least another century, explores how the interdependent 'organs' of society (people and institutions in their various roles and activities) are consciously or unconsciously involved in serving the needs and maintaining the order of society as a whole. This approach lends itself to a conservative outlook: functionalism can imply a level of social unity that does not exist. Moreover, purported holism has on occasion given rise to pernicious ideologies that have had disastrous consequences. Nazism, for example, strove to justify a rigidly authoritarian and 'racially pure' social order by proposing that, like any organism, society has 'parts' which exist *solely* in order to maintain the health and purity of the whole; the Nazi state and 'the *Volk*' (Harrington 1999).

In contrast to these examples, which can be considered 'conservative social holism', is the tradition that I have identified and dubbed 'radical social holism', or, more simply – 'radical holism'. The radical holists have adopted various holistic approaches (such as organismic biology, ecology and systems theory and even cybernetics) upon which they based their case for non-authoritarian, participatory, self-organised, humanly-scaled and decentralised social forms. Radical holism is a tradition which can arguably be dated from the mid-nineteenth century and that continued throughout the twentieth century. Its members would have variously described themselves as anti-authoritarian socialists, anarchists, communalists, social ecologists, or possibly none of these. More important than their different monikers is the connection they all made between the emancipatory social forms they advocated and form in the natural world. Preeminent in this tradition were the social ecologist Murray Bookchin (1922–2006) and the historian Lewis Mumford (1895–1990). Bookchin looked to the complementary, non-hierarchical patterns of relatedness found in ecosystems as the basis for an ethics upon which new kinds of human communities could be grounded (Bookchin 1980, 1982, 1986, pp.77–104). He contended that 'Either we will create an ecotopia based on ecological principles or we will go under as a species' (1980 pp.70–71) Similarly, Mumford argued that the 'The Organic World Picture' (with its portrayal of nature as a dynamically creative, yet stable and self-regulating realm)'undermined the conceptual framework of the dominant power system' (Mumford 1971, p.384 and p.391). The existential philosopher Martin Buber (1878–1965) summed up the position of many in this tradition in his call for 'a new organic commonwealth ... a community of communities' (Buber 1958, p.136) through which social structure could be renewed.

Whole systems science and philosophic holism

Both conservative and radical holism derive from the objective of social solidarity and yet the former leads to a defence of hierarchical, top down, social forms whilst the latter challenges these. It is paradoxical that the organic metaphor can be used in such divergent and contradictory ways. I propose that the problem lies in social theorists' incomplete and incoherent understanding of the nature of 'wholeness' and organic dynamics which is a result of mechanistic habits of thought. In recent years, however, there have been developments in science and philosophy which reinforce the radical holist case for forging a connection between natural form and liberated social form. Murray Bookchin's contention that 'Nature is writing its own nature philosophy and ethics' (Bookchin 2005, p.455) is supported by discoveries in chaos and complexity theory and a renaissance in the scientific approach to understanding the wholeness of natural organisms developed by poet-scientist J.W von Goethe (1749–1842) (see Miller 1988). The Goethean approach and chaos and complexity theory represent two complementary ways of thinking about nature, respectively the phenomenological/intuitive and the analytical.

One of the most important contributions to the Goethean renaissance has been made by philosopher and physicist Henri Bortoft in his book *The Wholeness of Nature* (Bortoft 1996), in which he makes an important distinction between 'authentic' and 'counterfeit' holism. This distinction, I will argue, is of the utmost importance to the development of a framework for the transition to a sustainable society.

Bortoft argues that wholeness is 'authentic' only when the whole form of a plant, an animal or even a text, is seen to emerge in and through its parts as they develop, over time. Each leaf of a plant, each organ of an animal, each word of a text, reveals a different aspect of the whole plant, animal or text. So, the whole can be said to be present or immanent in its parts; it is not 'a thing' that somehow exists separately from them, nor is it a sum of their total. Rather 'wholeness' is an experience, in the mind of the (Goethean) scientist or the text's reader, of the unity (or meaning) of the animal, plant or text. This experience is achieved through an encounter with the parts – organs, leaves or words. These parts become meaningfully/intrinsically related to one another through their mutual participation in the *coming into being, over time, of the whole*. Bortoft calls this dynamic mode of relationship *'belonging* together' (emphasis on the word 'belonging') (Bortoft 1998, pp.59–60): the parts *belong* together because they diversely express a single unity. Therefore, diversity is intrinsic to wholeness; the greater the diversity of *meaningfully related parts* that arise over time, the more fully the wholeness of a particular plant, animal or text can be realised.

Yet because of our sequential, linear habits of thinking (we think of whole *or* part rather than whole *and* part, one coming before the other rather than each mutually constituting one another) we do not usually think of wholes and parts as reciprocal. Those who are holistically inclined tend to give priority to wholes *over* parts, seeking wholeness, or unity, by taking *'a multiplicity of different things, and [subtracting] from them all the respects in which they are different to leave what they have in common'* (Bortoft 1999, p.93). 'Wholeness' is supposedly what remains after this stripping away process has occurred; the parts of such wholes are extrinsically rather than intrinsically related (they do not mutually participate in bringing forth the whole, they do not *belong* together) and therefore the unity of the whole is imposed or artificial. It is a way of understanding wholeness that Bortoft refers to as 'counterfeit' and it represents a mathematical, abstract, homogenising and static style of thinking, in which *'the universal is the authority and the particular simply does as it is told'* (Bortoft 1999, p.93). But, particularly since the Enlightenment, it is a style of thinking that has been extended into many fields – for example, biological classification, and ethics and aesthetics – in which 'universals' are sought (Bortoft 1999, pp.93–94). Moreover, as I will argue below, it is a form of wholeness that has come to be embodied in many of our institutions, with dire consequences.

Several of the key themes of systems theory (Von Bertalanffy 1968, Laszlo 1996) and its progeny, chaos and complexity theory (Capra 1996, Briggs & Peat 1989) can be understood in terms of 'the whole being present in parts'. For example, in the early to mid-twentieth century, systems theorists realised that whole systems in nature are always nested structurally – cells within organs, organs within organisms, organisms within ecosystems, and so on (see Figure 1). This arrangement is often referred to as 'a hierarchy' but this is a misnomer since any system at a given level of scale depends upon systems at lesser levels of scale in order to constitute itself: there is no forest without the trees and other organisms of which it is comprised, but there are no trees and other organisms without the organs and cells of which these are comprised. In other words, in nature's nested systems there is an interdependence of whole and part at all levels of scale and each level of scale is at once an whole in its own right, as well as a part that expresses an aspect of a greater whole. The scientist and novelist Arthur Koestler coined the term 'holarchy' to describe this relational nested structure (Koestler 1975, pp.45–70 and pp.314–48). Similarly, the themes of self-organisation and emergence, which are intrinsic to natural systems, describe the coordinated activity of the interrelated parts of a system which mutually constitute the whole to which they belong. Self-organisation could be defined as *the participation of each of the parts in the emergence of the whole*. The ongoing activities of thousands of different organisms in an ecosystem collectively and spontaneously give rise to the whole ecosystem.

Thinking in terms of the whole in the part, therefore, enables us to begin to integrate the Goethean and the systems/complexity approach to understanding nature, and supplies an expanded definition of what, as I said above, Bortoft refers to as 'authentic holism': to summarise, the authentic whole is present in and reliant on intrinsically related parts to come into being; it fosters diversity and participation; it is creative, self-organising, emergent and nested at different levels of scale. This contrasts with counterfeit holism in which whole and part are disassociated and in which, therefore, emergence, self-organisation, participation, relatedness, diversity and nestedness are all negated.

A new and radical social holism

The distinction between counterfeit and authentic holism is fundamental to the development of a holistic framework for transition to a sustainable society. There is an affinity between counterfeit holism and top down/ authoritarian social forms: when translated into the realm of human affairs, giving 'the whole' priority over the parts serves to legitimate control by supposed (i.e. counterfeit) social 'wholes' over the individuals (the social 'parts') of whom they are comprised. This is not just a theoretical position: there have been numerous historical occasions, from Plato to the Nazis (Popper 2002, Harrington 1999, Marshall 1992 p.407–8) when ruling classes have declared that their populace has an obligation

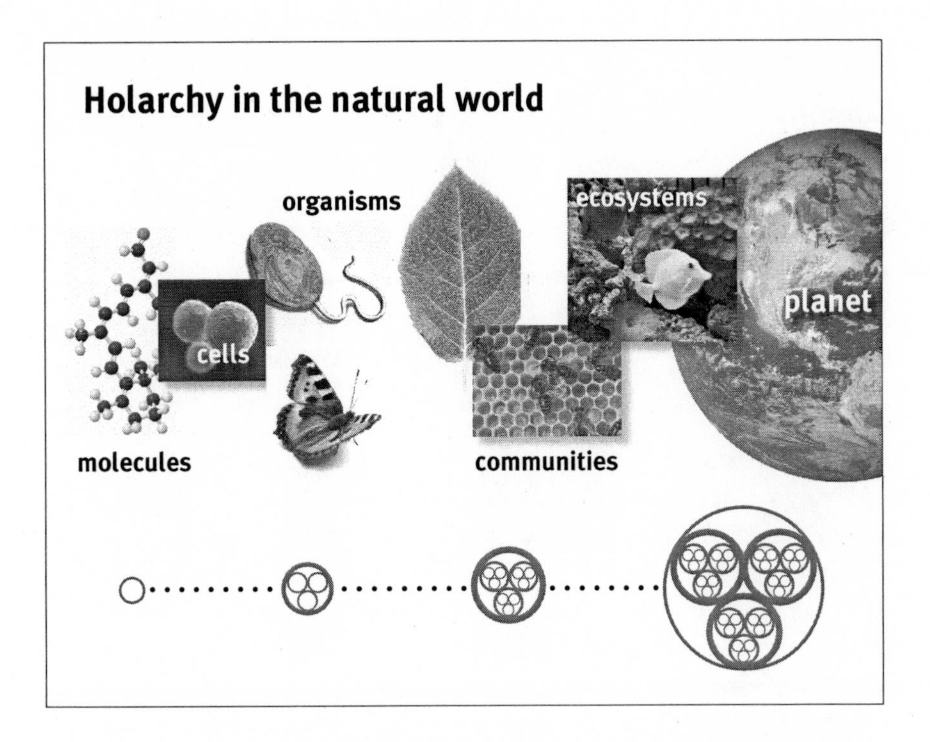

Figure 1. Holarchy in the Natural World. Natural forms are arranged in nested 'holarchies' of whole/parts, or 'holons'. Each such holon is at once a whole in its own right, but a part of a greater whole and is therefore semi-autonomous or interdependent with other holons, as well as self-organising and emergent. Just as holarchies are central to the sustenance of natural form, so too holarchies of households, neighbourhoods, villages, towns, cities and regions are key to establishing sustainable patterns of everyday life. Diagram by Terry Irwin.

of unquestioning allegiance/obedience to the *greater whole* (the city state, the nation state, 'the *Volk*', to name but a few examples of supposed social wholes) and often this greater whole is explicitly compared to an organism. It was for this reason that many liberal philosophers such as Karl Popper rejected social holism out of hand (Popper 2002).

Similarly, there is a convergence between what I have defined as the radical holist tradition and authentic holism: the themes of self-organisation, participation and mutualistic relatedness are central to both. However, radical holists did not have the distinction of authentic and counterfeit holism, and were therefore unable to adequately refute the charge that holism is inherently authoritarian, or understand how to more usefully apply the organic metaphor to the social realm. This accounts for Murray Bookchin's ambivalence about the concept of 'social wholeness':

historically, he contended, it has been sought *'through homogenisation, standardisation and a repressive coordination of human beings'* (Bookchin 2005 p.88). But the distinction between counterfeit and authentic holism establishes a more robust connection between the concept of holism and emancipatory social forms than has been hitherto possible. There is also, as I will argue below, an intimate association between authentic holism and the sustainability of everyday life.

Context: everyday life and needs

I began this chapter by referring to a long list of problems that affect every aspect of our lives and that of other species, and argued that a framework was necessary to assist the process of problem framing and solving that will move us in the direction of a sustainable society. The context within which problems arise and solutions are to be developed is everyday life: it is the foundational level for all human experience, and we are unavoidably immersed in it. A framework for transition, therefore, needs to be embedded in everyday life.

To begin, we must start in a realm that is even more fundamental than everyday life – that of human needs. Everyday life is brought into being as people strive to satisfy their needs. Development economist Manfred Max-Neef and colleagues (Max-Neef *et al.* 1991) have developed a theory of what are claimed to be ten material and non-material needs that are common to all cultures: subsistence; affection; participation; creation; understanding; identity; freedom; protection; idleness; transcendence. (See Table 1.) There is room for debate over the structure and details of this taxonomy (for example, whether a given need is 'universal'). However, the more important point, which differentiates it from all other theories of needs, is the distinction that is made between *needs* and the *means by which needs are satisfied*: whilst needs are universal, 'satisfiers' vary wildly from culture to culture and place to place and from one historical period to another.

It is the variation in how needs are satisfied that gives rise to the diversity of forms of everyday life that have arisen all over the planet, and that make everyday life *specific to place.* To take a simple example, one community may satisfy its food needs (i.e. part of its subsistence needs) by fishing, another by farming, and another by hunting. The respective differences in the means by which needs are satisfied is one of the reasons

Needs (universal)	Examples of satisfiers (unique to time/ place/ culture)
Subsistence	Food, shelter, clothing
Participation	Associations, churches, councils
Protection	Healthcare, shelter, social security
Affection	Friendship, family
Creation	Workshops, cultural groups, craft, music
Understanding	Literature, education, meditation
Identity	Customs, traditions
Freedom	Political organisations, councils
Idleness	Games, parties, sun bathing
Transcendence	Meditation, religion, spiritual practices

Table 1: A simplified rendition of Max-Neef et al.'s matrix of needs and related satisfiers (Max-Neef et al. 1991, pp.32–36). Everyday life is shaped according to how the needs in the left column are satisfied by the activities in the right column. Some satisfiers will simultaneously satisfy multiple needs.

that the everyday life in fishing/farming/hunting communities is so different. Furthermore, depending upon whether control of the satisfaction of needs is internal (endogenous) or external (exogenous) to a community, two fundamentally different forms of everyday life arise. In the former case, everyday life comes to embody many of the features of 'authentic wholes', described above; in the latter case, everyday life comes to embody many of the features of counterfeit wholes.

Control over the satisfaction of needs from within communities and the concept of authentic wholeness in everyday life are two sides of the same coin. If a community strives to satisfy its subsistence needs, members collectively plan, manage, grow/hunt/fish, harvest, preserve, process, store, and prepare their food. To satisfy their needs (whether material or non-material) individuals must establish collaborative/mutualistic forms of interrelatedness through which their communities are self-organised. To the extent that they do this everyday life within communities will be emergent; it will not be subject to imposed blueprints or top down control. Just as the diverse parts of a plant mutually participate in bringing forth the whole plant, each expressing a different aspect of its wholeness (and by virtue of this mutual participation *belong* together), so when needs are endogenously satisfied (when the satisfaction of a given need is controlled from within the community in which it arises) individuals mutually participate in bringing forth everyday life. In this way, the diverse 'parts' of everyday life (not only people, but also the natural world and artifacts) come to *belong* together, or form an organic-like unity. In the next section I shall discuss the exact form that wholeness takes in everyday life, but suffice it to say this kind of wholeness must lie at the heart of transition.

To the extent that a community voluntarily or involuntarily cedes control of the satisfaction of its needs to external and centralised institutions, everyday life will lose the aforementioned qualities of wholeness. For example, if food is grown outside of a community's boundaries and is distributed and sold by shareholder owned companies (the agenda of which is very different to the community in question) then the facet of everyday life that is represented by the satisfaction of the subsistence need is now externally controlled. In such an example the satisfaction of the need is managed rather than self-organised and is directed rather than emergent: the decline of self-organisation and emergence represents a decline in the community's freedom. The mutual participation that previously enabled people, artifacts and nature – the parts of everyday life – to *belong* together is now diminished. As *belonging* together diminishes, so too does everyday life's vitality, reciprocity and creativity. There is a direct correlation, therefore, between the extent that 'satisfiers' are taken out of a community's control and development of alienated relationships between people, their artifacts and nature, the fragmentation of everyday life. All of the problems I begun this chapter by outlining can be described in terms of this three-way alienation process. *There is, then, a causal relationship between the loss of control of the satisfaction of needs and the unsustainability of everyday life.*

Just as authentic wholeness and control of the satisfaction of needs are two sides of the same coin, so too are loss of control of the satisfaction of needs and counterfeit holism. Just as authentic wholeness is at the heart of the framework for transition, counterfeit wholeness, with its association with the loss of social cohesion, cultural diversity, community autonomy and a harmonious relationship with the natural world, is at the root of many of the problems I outlined at the beginning of this chapter.

The Domains of Everyday Life

I have discussed how everyday life arises out of the satisfaction of needs, and how two fundamentally different patterns of everyday life are created according to whether communities are in control of this process: self-organised vs. managed, emergent vs. top down, participatory vs. non-participatory. But whilst I have spoken in general terms about these contrasting forms of everyday life I have not been specific about the form that wholeness takes in everyday life. The question remains: if the *parts* of everyday life are people, the natural environment and artifacts, what are the *wholes* within which self-organised and participatory activities take place?

When communities are in control of the satisfaction of their needs, everyday life takes on a nested structure: households exist within villages and neighbourhoods, villages and neighbourhoods within towns or cities, towns or cities within regions. I have coined the term Domains of Everyday Life to describe these social forms (see Figure 2). When these are vital, they are the structural *wholes* of everyday life; nested, bounded and networked forms, the emergence of which reflects the place-specific ways in which the participatory and self-organised satisfaction of needs occurs. These Domains are semi-autonomous and mutually interdependent wholes, since needs can only be partially satisfied in any given Domain. The inhabitants of a neighbourhood, for example, will still depend upon other neighbourhoods and the city in which their neighbourhood is ensconced for the satisfaction of many of their needs. The integrity of the Domains, the vitality of everyday life within them, is related to the degree to which needs are satisfied through self-organisation and participation, which in turn is dependent on the quality of relatedness within and between the Domains. In short, the Domains provide the context for everyday life and, when they are vital, their structure resembles an ecosystem or other organic system.

A cursory glance at descriptions of pre-industrial capitalist communities, or communities only recently affected by this process, demonstrates the range of needs that were once typically satisfied within the Domains, giving them a great deal of vitality. For example, the historian Kirkpatrick Sale lists fourteen different kinds of artisans (plus shopkeepers, publicans, and local farmers) that would have served a typical small English town of about two thousand people even in the late nineteenth century (Sale 1980, p.400). The chronicler of rural life

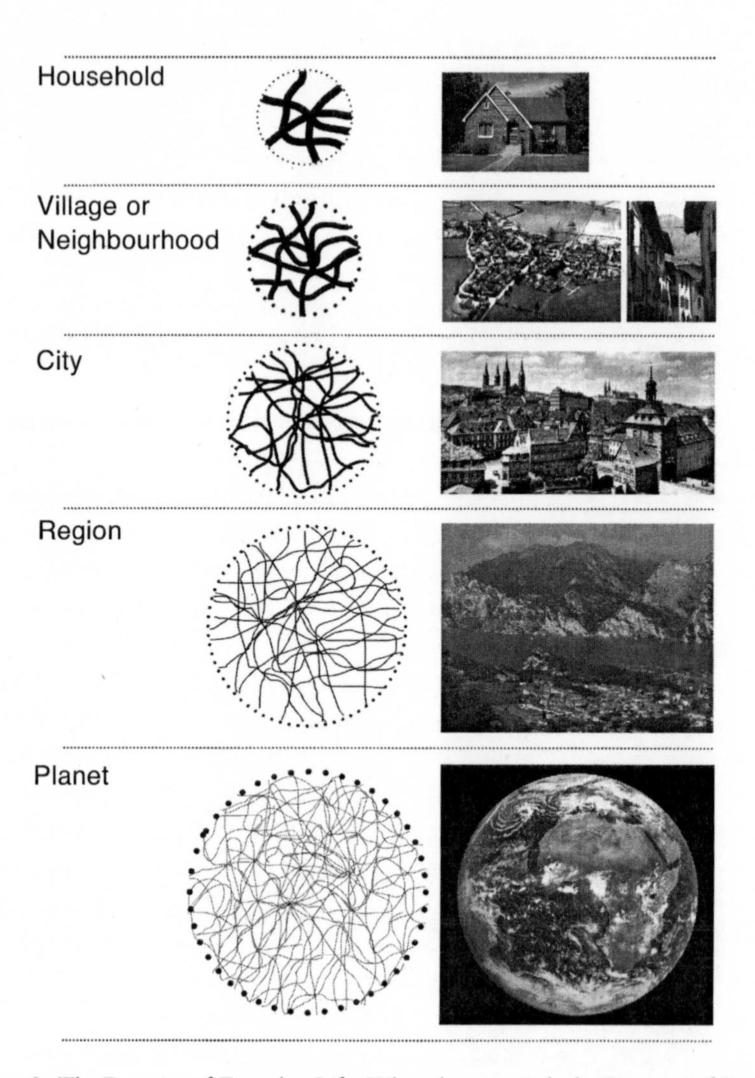

Figure 2. The Domains of Everyday Life. When they are vital, the Domains of Everyday Life represent different kinds of community, each with its own typical characteristics. This is a reflection of the different patterns of relationship – between people, the things they make and nature – that are necessary to satisfy needs at different levels of scale. Different kinds of activities are therefore appropriate to each Domain. Diagram by Terry Irwin.

Norman Wymer lists a similar number for a much smaller English village (Wymer 1951, p.37). Many anthropological studies of non-westernised societies also reveal the vitality of everyday life at each level of scale. For example, in anthropologist Helena Norberg-Hodge's study of Ladakh, *Ancient Futures* (2000), it is possible to discern how, as recently as the

1970s, everyday life in household, village, and region are self-organised around extensive collaborative networks. Through these, not only were needs satisfied in many different ways, but also Ladakhis came to *belong* together, that is, they created unified communities which harmonised with the natural world in which they were embedded.

When the satisfaction of needs is controlled by communities *in place*, each of the Domains of Everyday Life assumes its own unique role in the life of the community: the same needs may be satisfied, but in different ways at each level of scale. For example, at the level of the household, the need for 'affection' would be satisfied by long term, often biologically based, multigenerational relationships, whilst at the level of the neighbourhood, the same need would be satisfied by more freely chosen friendships. Similarly, the same need is likely to be satisfied in different ways at the same level of scale in different places. So, although the Domains of Everyday Life could be described as universal/archetypal forms that have been common worldwide, throughout history, these archetypes have been diversely expressed as emergent properties of people satisfying their needs in ways appropriate to their *time, culture and place*.

When the Domains are vital, each represents a different level and facet of community, and each has its own qualities and possibilities. Moving from inner Domains to outer Domains, from the household through to the region, relationships between people, their artifacts and nature become progressively less intimate and more transient, but more multiple and diverse. When many needs are satisfied within the household (as was the case in many pre-industrial communities) it could be seen as a small, tightly bound 'community' based on relatively few, long-term relationships; under similar circumstances the Domain of the Neighbourhood is a larger, less tightly bound community, but has more variegated relationships, and so on, moving out through the Domains. It is this shift from 'thick' to 'thin' and few to many relationships that accounts for the changing character of everyday life at different levels of scale. For example, in pre-industrial society the Domains of the Household, Neighbourhood and Village were levels of everyday life better suited to the creation of livelihoods than the Domain of the City, which provided a market that enabled neighbourhoods, villages and households to trade. The Domain of the City or Region was more likely to support universities, hospitals and other cultural institutions than the inner domains.

When self-organisation, participation and interrelatedness are highly developed, each of the 'parts' of everyday life contributes to the emergence of the wholes of everyday life – the Domains. Take for example, the activity of making and eating a loaf of bread baked in a household, placed on a board, and cut with a knife. This represents an interaction of the human, the natural and the artifactual that together help the household emerge as a self-organised form. When the bread is made from wheat grown in the local countryside and processed in the city; the board is made from wood from local forests and crafted in a nearby village; the knife is engineered in a local workshop, then baking and eating a loaf in the household also helps region, city and neighbourhood emerge as self-organised forms, and bread, board and knife all express different but related aspects of the household, neighbourhood, city and region. In short, bread, knife, board, and the people who are making and using these, *belong* together: everyday life has organic unity.

There are very few, if any, pre-industrial societies in which everyday life could be described as purely and unambiguously authentically 'whole'; elements of control, coercion, stratification and fragmentation are usually present. Nevertheless, most pre-industrial societies managed to retain something of this basic organic form and managed to live sustainably 'in place' for generations. Described in these terms, we can begin to see what is meant by the loss of 'organic' social form that has been repeatedly bemoaned by many sociologists, historians, anthropologists and philosophers. The sociologist Gerard Delanty notes, for example, the *'discourse of loss'* that is at the heart of *'modern thought from the Enlightenment onwards ... with a sense of the passing of an allegedly organic world'* (Delanty 2003, p.11).

The decline of the Domains of Everyday Life

As social and economic theorist Mario Kamenetzky notes, control of the satisfaction of human needs by élites has been one of the hallmarks of civilisation (Kamenetzky 1992, p.186), and is a reflection of its hierarchical structure. In our era this phenomenon has penetrated so far into everyday life that satisfiers are predominantly externally controlled. This has resulted in the demise of the Domains of Everyday Life. They are still present – we still have 'households', 'villages, 'neighbourhoods', 'cities' and 'regions' – but instead of functioning as semi-autonomous and robust wholes that integrate the satisfaction of needs in everyday life, today the Domains have become vestigial; they are fragments of externally controlled, unsustainable globalised systems run by institutions that are unaccountable to the communities they service. Typical of the radical holists was Martin Buber's observation that society has been *'hollowed out'* by industrial-capitalism (Buber 1958, p.14). Conceptualising everyday life in terms of the Domains makes it clearer *what it is* that has been hollowed out, and how this has occurred.

The result is that for most people on the planet, subsistence needs can now only be met by engaging in the globalised market place, which is centrally controlled by a handful of corporations. The primary objective of these is not to satisfy needs but to make a profit. The non-material needs identified by Max-Neef, such as 'understanding', 'participation', 'freedom' and 'security' have suffered a similar fate, as the political process, education, leisure, healthcare and so on have been appropriated by both the market and the nation-state.

The decline of the Domains, or what could be considered ecosystems of everyday life, parallels the destruction of natural ecosystems, and the consequences of their decline are many, and global in scope. Many of the problems that I began this chapter by listing are expressions of the erosion of the organic structure of everyday life. For example, social alienation, and any number of related issues, arises out of the decline in community at every level of scale; regional culture disappears as unique place based satisfiers controlled from within the Domains are abandoned; endlessly sprawling megalopolises reflect the disappearance of the boundaries within which communities once constituted themselves. In other words, as the Domains of Everyday Life go into decline, society becomes ecologically, socially, economically, politically and culturally unsustainable.

No institution or social phenomenon is entirely counterfeit or authentically whole, elements of each are no doubt always present. Therefore it would be an oversimplification to portray everyday life in pre-industrial communities as an unalloyed expression of authentic wholeness; and it is a mistake to represent contemporary everyday life as an unalloyed expression counterfeit wholeness; it is always a matter of degree. Nevertheless, whereas formerly the Domains were the context within which needs both arose and were satisfied, now this intimate association has been eroded: everyday life in modern society is presided over by institutions (such as central and local government, corporations of all kinds, universities and schools, hospitals, and so on) that are dedicated both to the production and control of satisfiers. This results in the extreme centralisation of everyday life, and a situation in which many needs are satisfied inadequately and in a piecemeal fashion.

For example, many historians and anthropologists have shown that in pre-industrial communities, harvest of food not only met the need for subsistence, but was also a festive occasion which involved the whole community: several needs identified by Max-Neef – subsistence, participation, affection, idleness – were all satisfied simultaneously in an integrated way. The harvesting of food by a large corporation, on the other hand, is aimed at satisfying a single need alone (other institutions are relied on to satisfy the other needs) and in as far as the food produced is toxic and nutrient poor this need is not satisfied adequately. The profit generated from this harvest is funnelled to shareholders and highly placed individuals within the corporation to enable these individuals, to paraphrase Mario Kamenetsky, to acquire more than their fair share of satisfiers (Kamenetzky 1992, pp.185–86). Whilst within such a corporation there will be elements of self-organisation and participation, such a corporation is, structurally speaking, an overwhelmingly counterfeit whole, hierarchically organised and non-participatory. Because of this, the quality of relationships within such organisations is low, and top-down management is necessary in lieu of any natural social unity.

The process of appropriation of decentralised and place-based satisfiers by counterfeit social wholes eviscerates the Domains. Counterfeit holism is essentially a process of homogenisation and as everyday life is increasingly dominated by organisations that can be thought of as counterfeit wholes, so everyday life itself becomes homogenised. The story is the same the world over: it is the story of industrial-capitalist civilisation's encounter with a pattern of everyday life as old as humanity itself. The diversity found in the myriad manifestations of the household, village, neighbourhood, city and region disappears in the face of a globalised but fragmented homogeneity that is controlled by institutions that are not meaningfully integrated into the web of relatedness of any Domain.

The Domains of Everyday Life and the transition to a sustainable society

The transition to a sustainable society will require the reconstitution and reinvention of households, villages, neighbourhoods, towns, cities and regions everywhere on the planet as interdependent, nested, self-organised, participatory and diversified wholes. This will essentially be the transition from counterfeit to authentic holism in everyday life. The result will be a decentralised and diversified structure of everyday life which is in contrast to the centralised and increasingly homogenised structure that we have become accustomed to. It will resemble the *'community of communities'* that Martin Buber envisioned (Buber 1958, p.136), except that it will embody the communion not just of people, but of people, their artifacts and nature, and will come into being at multiple, interrelated levels of scale.

It should be emphasised that this should not, and cannot, represent a simple return to traditional life ways. Modernity has brought with it many social and technological advances that should not be dispensed with. Furthermore, an additional Domain, the Domain of the Planet, has been introduced. This Domain can potentially bestow upon everyday life a cosmopolitanism and diversity which was not available to pre-industrial communities. Reconstituting the Domains, therefore, is not simply about re-localisation; it is about establishing a symbiotic relationship between the global and the local.

The Domains cannot be reduced to separate social forms, economic forms, political forms, cultural forms, technological forms, artistic forms or architectural forms: when they are vital they represent the integration of all of these facets of everyday life *in ways unique to particular places by particular communities*. Reconstituting the Domains is an inherently transdisciplinary and grassroots process that represents an opportunity to reintegrate and recontextualise knowledge, embedding it in both community and everyday life. It calls for the intentional, or designed, reintegration of all facets of everyday life in place, and suggests that a new kind of designer is needed, a 'transition designer'.

Any place-based self-organised and participatory activity aimed at satisfying a need at any level of scale, that protects or engenders the intrinsic relatedness (the relatedness of people, artifacts and nature) within and among the Domains, will contribute to the transition to a sustainable society.

Such activities will comprise facets of everyday life through which the Domains will reemerge as vital, semi-autonomous authentic wholes. Firstly, we need to look at the Domains, as they exist today, and ask what needs are currently being endogenously satisfied, and how this satisfaction might be protected and 'amplified', to coin sustainability designer Ezio Manzini's term (Mendoza 2010). Secondly, we need to develop new, endogenous satisfiers for needs that are currently exogenously (and probably inadequately) 'satisfied'.

The variety of endogenously organised and controlled projects and practices that could be established (or where they are already present, must be protected) within this framework are as limitless as the myriad features of everyday life itself. When projects and initiatives begin to be connected and integrated in particular places (the farm and forest with the market, the cafe, the grocery, the health centre, the garden, the larder and the composting toilet; the workshop with the laundry, the cinema, the factory, the transport system and the renewable energy facility; the school, the bank, the art studio, the councils with all of these) they will create ecosystems of interdependence and mutual benefit, parts and wholes of everyday life at all levels of scale enfolding and reciprocating one another.

Conclusion

I began this chapter by arguing that a framework for transition needed to provide a narrative which explains how our contemporary plight arose; a vision of a desirable alternative to contemporary society; a means of addressing problems and connecting solutions at appropriate levels of scale; a structure within which transdisciplinary and grassroots collaboration can take place; and a humane definition of sustainability. In updating the radical holist tradition by the application of a holistic paradigm to everyday life, all of these features of a framework have emerged. Hopefully it will provide a useful tool to 'transitionists', whatever problem their efforts are directed at, and wherever they are on the planet.

References

Biehl, Janet & Bookchin, Murray (1998) *The Politics of Social Ecology*, Black Rose, Montreal.

Bookchin, Murray (1980) *Toward an Ecological Society*, Black Rose, Montreal.

—, (1986) 'Ecology and Revolutionary Thought' in *Post-Scarcity Anarchism*, Black Rose, Montreal.

—, (2005) *The Ecology of Freedom*, AK, Edinburgh.

Bortoft, Henri (1996) *The Wholeness of Nature*, Floris Books, Edinburgh.

—, (1999) 'Goethe's Organic Vision' in *Wider Horizons* ed. Lorimer, David, Clarke, Chris, Cosh, John, Payne, Max & Mayne, Alan, Scientific and Medical Network, Leven, Scotland.

Briggs, John & Peat, David F. (1989) *Turbulent Mirror*, Harper & Row, New York

Brown, Richard Harvey (1989) *A Poetic for Sociology*, University of Chicago Press, Chicago.

Capra, Fritjof (1996) *The Web of Life*, Harper Collins, London.

Delanty, Gerard (2003) *Community*, Routledge, London.

Kamenetsky, Mario (1992) 'Human Needs and Aspirations' in *Real-Life Economics*, ed. Ekins, Paul & Max-Neef, Manfred, Routledge, London.

Elden, Stuart, Lebas, Elizabeth & Kofman, Eleonore (2003) *Henri Lefebvre, Key Writings*, Continuum, New York.

Escobar, Arturo (2009) 'Another World is (Already) Possible,' *World Social Forum, Challenging Empires*, eds. Sen, Jai & Waterman, Peter, Black Rose, Montreal.

Gardiner, Michael E. (2000) *Critiques of Everyday Life*, Routledge, London.

Gordon, Scott (1991) *The History and Philosophy of Social Science*, Routledge, London.

Hawken, Paul, Lovins, A. & Lovins, L. Hunter (1999) *Natural Capitalism*, Little Brown & Co, London, UK.

Harrington, Anne (1999) *Reenchanted Science*, Princeton University Press, Princeton.

Highmore, Ben (2002) *The Everyday Life Reader*, Routledge, London.

Hollis, Martin (1994) *The Philosophy of Social Science*, Cambridge University Press, Cambridge.

Hopkins, Rob (2008) *The Transition Handbook*, Green Books, Dartington, UK.

Jacoby, Russell (2005) *Picture Imperfect*, Columbia University Press, New York.

Koestler, Arthur (1975) *The Ghost in the Machine*, Pan Books, London.

Laszlo, Ervin (1996) *The Systems View of the World*, Hampton Press, Cresskill.

Marshall, Peter (1992) *Nature's Web*, Simon & Schuster, London.

Max-Neef, Manfred A. *et al.* (1991) *Human-Scale Development*, Apex Press, New York.

Mendoza, Andrea (2010) *Desis Newsletter 2: Amplify Workshop.* At http://www.desis-network.org/?q=node/356

Merchant, Carolyn (1983) *The Death of Nature*, Harper & Row, San Francisco.

Miller, Douglas (1988) *Goethe, Scientific Studies,* Princeton University Press, Princeton.

Mumford, Lewis (1961) *The City in History,* Harcourt, Brace & World, New York.

—, (1971) *The Pentagon of Power*, Secker & Warburg, London.

Norberg-Hodge, Helena (2000) *Ancient Futures*, Rider, London.

Phillips, D.C. (1976) *Holistic Thought in Social Science,* Stanford University Press, Stanford.

Polkinghorne, Donald (1983) *Methodology for the Human Sciences,* State University of New York Press, Albany.

Popper, Karl (2002) *The Poverty of Historicism*, Routledge, London.

Sale, Kirkpatrick (1980) *Human Scale,* Secker & Warburg, London.

Stark, Werner (1962) *The Fundamental Forms of Social Thought,* Routledge & Kegan Paul, London.

Von Bertalanffy, Ludwig (1968) *General System Theory*, Penguin, Harmondsworth.

Walker, Brian & Salt, David (2006) *Resilience Thinking,* Island Press, Washington.

Wymer, Norman (1951) *Village Life,* George G. Harrap, London.

12. Applied Storytelling

SHAUN CHAMBERLIN

Humanity has dramatically changed the world we live in, creating soil fertility depletion, fish depletion, fresh water depletion, climate change, ocean acidification, peak oil, chemical pollution, biodiversity devastation, inequity and war, among other crises.

Other chapters in this book propose solutions to many of these, but we must also admit that some are not mere problems, but genuine predicaments – circumstances which cannot be changed, but only adapted to. Just as the inevitability of one's own death cannot be altered, so there is no way we can prevent the climate change already 'locked in' to the system, restore the species already lost, or unburn our fossil fuels.

Indeed, our recent history might be described as a process of slowly turning soluble problems into predicaments, one by one.

But that is not to say that we do not face important, even critical, decisions. Many of our problems can still be solved, and even life's predicaments must be dealt with. Even in the most difficult of times, there are better and worse courses of action.

As NASA's Dr James Hansen, probably the world's leading climate scientist, has concluded, our destabilisation of the Earth's climate has now reached the point where there is only one remaining hope for preserving planetary conditions similar to those for which life is adapted (Hansen 2008). Humanity itself must consciously act to produce a deliberate cooling force, to counteract our own former actions.

In other words, it is clear that Mother Earth is no longer able to quietly clean up after us – it is time for our childish culture to grow up, and either take some responsibility for our actions or face the consequences. As the Chinese proverb has it:

> If you don't change direction, you are likely to end up where you're headed.

But changing direction is not a simple matter. Of course there are myriad practical challenges to overcome, but humanity has proved itself adept at that. The more fundamental requirement is a change in our perspective, *in our stories.*

In human cultures around the world and throughout history, stories have told us what is important, defined our identity and shaped our perceptions and thoughts. This is why we use fairy stories to educate our children, why advertisers pay such extraordinary sums to present their creations, and why politicians present both positive and negative visions and narratives to win our votes. We choose our actions on the basis of our stories. It only takes an hour to learn how to plant a tree, but it might take a lifetime to learn why you would want to.

In our culture, the story that humanity exists only to consume the fruits of a world laid out for our convenience retains great influence, and other powerful stories guide us too. The dominance of the story of 'progress' – that we currently live in the most advanced civilisation the world has ever known, and that we are advancing further and faster all the time – makes 'business as usual' an attractive prospect. Why would we not wish to continue this astonishing advancement?

And, perhaps most influential of all, our governments entrust the decisions that shape our lives and futures to economic theories which tell stories of the 'invisible hand' of the market, which calmly allocates resources and energy in the best possible way through the divine conjunction of competition, supply and demand and self-interest.

It may seem unbelievable that these stories still hold sway in the face of the ever more damning scientific evidence of the impacts our choices are having on the world, but psychologists have long known that our stories are highly resistant to facts that challenge our worldview and identity (see, for example, Sherman & Cohen 2006, pp.3–6). We have so much invested in these stories that it is as though our very self-worth is under threat ('Is my chosen lifestyle, or my belief-system, really contributing to the death of our world?'), and we have an arsenal of psychological responses to such threats, reinforced by the relentless, overwhelming *normality* of the mainstream culture all around us.

Of course, viewed historically, our modern way of life is anything but normal, but for many of us it is all we have known, and even if we are not comfortable with it, the powerful cultural story that 'real change is impossible' (despite all evidence to the contrary) urges us to accept things as they are.

We must also look at those with the most power and influence. Those who benefit most from the current 'business as usual' arrangements in the global economy are, by definition, extremely wealthy, and thus very influential. They have disproportionate access to the mass media, and are highly respected. And it may be these people for whom it is hardest to acknowledge the reality of our collective crisis.

Everything and everyone around them tells them that they have made an incredible success of their lives. Their ability to acquire whatever material possessions they desire, their exclusive social groups, their desirable romantic partners and respectful business colleagues – so much in their lives speaks to them of their triumphant prosperity.

Quite apart from the challenge of changing the habits of thought and reasoning that have appeared to serve them so well, what we are asking is that they shift the story they hold of their own life from one of heroic progress and success to one in which they have been complicit in the ongoing destruction of much that is precious on Earth.

We should not underestimate the level of commitment and bravery such an internal transformation requires, especially when their peer groups provide them with all the encouragement in the world to resist it. And so we should not be surprised when many instead indulge in desperately contorted logic or straight-out denial (and bring their considerable resources to bear) in the attempt to retain the glorious tale of themselves that they toiled and sacrificed to earn.

Yet despite all this, there are those who have braved this dark night of the soul. In 1999, Ray Anderson, founder and chairman of vast multinational corporation Interface Carpets began a speech to fellow business leaders with the following words:

> Do I know you well enough to call you fellow plunderers? One day it dawned on me that the way I'd been running Interface Carpets is the way of the plunderer. Plundering something that is not mine, that belongs to every creature.
>
> I stand convicted by me, myself alone, not by anyone else, as a plunderer of Earth, but not by our civilisation's definition. By our civilisation's definition I'm a captain of industry, in the eyes of many a kind of modern-day hero. (Anderson 1999)

He deserves huge respect for embracing this difficult epiphany (he has likened the experience to 'the point of a spear into my chest'), and his task now, and ours, is to build new cultural stories, so that his 'fellow plunderers' may come to measure their success very differently, but again become heroes.

This seems to me to be the fundamental challenge we face – changing the stories that define success, identity and meaning in our culture.

Fortunately, there is a vast and diverse upwelling of people, organisations and communities who are acutely aware of the evidence painstakingly collected by our scientists, and are forging new stories that might better serve our collective future. Paul Hawken (Hawken 2008, p.141) has described this self-organising human response as being part of the Earth's immune system, working to counter the very real threat to life on the planet (to us).

The mainstream media chooses not to tell us much about these groups, preferring to champion consumerism, but where would any of us be if our own immune system got distracted seeking its personal fortune, say, or pursuing hedonistic diversions?

The truth is that the largest movement in the world has grown up without name or solid structure, simply as a response to very human desires for safe water and food, for justice, for diversity and for health. Above all, for a future.

As the political journalist and author Norman Cousins wrote, 'all things are possible once enough human beings realise that everything is at stake' (quoted in Hopkins 2008, p.140). This awareness is spreading fast.

The strand of this movement that I have been most involved with is the Transition Towns, a network of communities around the world (ranging from *favelas* in Brazil to Japanese towns, from rural villages in England to Transition Los Angeles!) unified in their drive to devise and implement positive, enjoyable solutions that build local resilience and reduce fossil fuel dependency. Transition takes to heart the words of one of its pioneers, my late colleague and friend Dr David Fleming:

> Localisation stands, at best, at the limits of practical possibility, but it has the decisive argument in its favour that there will be no alternative.
> (Quoted in Hopkins 2006.)

The practical manifestations of this movement are as diverse as the communities that give birth to them, ranging from food cooperatives, local currencies, skill-sharing sessions and renewable energy projects to Transition Universities, arts projects and Energy Descent Action Plans for entire regions, endorsed by local Councils.

But what excites me most is that these practical projects are growing out of a new sense of how we should respond to radically changing times. Our culture generally offers two ways of responding to crises – individual action like changing light bulbs or biking instead of driving, and political lobbying of those in power to get changes made. Yet individual action can feel insignificant in the face of global challenges, and political lobbying is often simply ignored. With these the only apparent options, it is easy to see why many become disheartened and return to going with the flow.

Transition offers a meaningful alternative, and I believe that this is part of the explanation for its ongoing rapid growth. Transition is local communities recognising that the dominant stories in our culture have no future, and grasping the opportunity to profoundly rethink much of what we have come to take for granted. To borrow a phrase from the actor and activist John O'Neal, it is throwing an anchor into the future we want to build, and pulling ourselves along by the chain; taking direct action on a more significant scale than can be achieved alone, yet where individual input remains valued and significant.

Perhaps most meaningful of all is the realisation that when communities decide to shape the future of their village, town or city, there is no authority that stops them. In fact, Transition communities often discover that their local Council have been sitting around having meetings about how to engage with the local public! Turning up to discuss change often involves pushing at an open door.

And by getting together to act, communities become communities again, rather than just groups of independent individuals. We learn to rely on each other again for some of our needs, rather than on our money and the complex systems of the global economy. We get to know each other's strengths and limitations, understand each other's characters, and learn to co-create. So many attempts at recapturing our lost 'sense of community' believe that getting together every Tuesday evening is enough, but it is not. The basis of community is not simply meeting up and being nice to each other, it is truly depending on each other.

I have been involved with the Transition movement for five years, as one of the co-founders of Transition Town Kingston, and as the author of the movement's second book *The Transition Timeline*, which develops the theme of Transition's role as storyteller.

The book was requested by existing Transition initiatives who were trying to produce realistic positive visions of the future for their communities, but needed some input on the major trends facing the UK in our near-future – the kinds of things which are going to affect them but which may be hard for individual communities themselves to directly affect (such as peak oil, government policy or UK food supply). Accordingly, the book provides a readable summary of the existing research in key areas of concern like population, food and water, energy and healthcare, outlining the present position and trends before exploring possible futures. It also details tried and tested techniques that Transition initiatives have found useful in their local planning and projects (Chamberlin 2009).

But the heart of the book is a fleshing out of what we called the 'Transition Vision'. This concept was touched on in Rob Hopkins' *Transition Handbook*, but Rob and I felt strongly that it needed to be developed further. There is a great deal of despair among people who see the state of our world, but I believe that motivation lurks hidden within despair. Despair is looking at the future we see as inevitable and realising how far it is from what we want. All that is required is a realistic vision of a desirable alternative, and despair immediately transforms itself into a massive motivation to action.

The Transition Vision, then, is of a future in which we create a resilient, more localised society which avoids the worst potential of climate change and peak oil through building thriving lower-energy communities teeming with satisfying lifestyles and fulfilled people. This vision was developed in collaboration with as many Transitioners and others as I could manage to speak to, and in the book this vision is tracked through a 'history of the next twenty years', setting it alongside three other possible futures based on different stories of our place in the world. Some of these might be justifiable grounds for despair, *if* they were our only option.

I believe that the Transition movement, with its new stories for community action, represents a powerful force for a better future, especially as part of the wider movement for ecological sustainability that is swelling all around us. But we must also recognise that, given the global nature of many of our problems, local solutions are necessary but not sufficient. Small-scale solutions can struggle to match up to large-scale problems. But David Fleming addressed these challenges of scale with a phrase that transformed my thinking on the matter, and has remained a touchstone for me ever since:

> Large-scale problems do not require large-scale solutions – they require small-scale solutions within a large-scale framework.
> (Fleming 2007, p.39.)

The truth is that large-scale solutions too face their own inherent problem – that of disconnectedness. For example, while it is tempting to think of hard-won political agreement on a tightening global cap on emissions as a solution to climate change, such a cap is meaningless without on-the-ground solutions at the local and individual levels. This is, after all, where those emissions are generated.

The true challenge lies not in the essential process of agreeing a cap, but in transforming our society so that it can thrive within this limit. If we fail in this, the pressure to loosen or abandon any cap will become irresistible – *'enough talk of future generations, my children are hungry today'*.

It is clear that we need the kind of global agreement on climate change put forward in schemes like Contraction and Convergence or Greenhouse Development Rights, which would assign clear national carbon budgets within an adequate global response. And it is clear that we need local responses of the kind that Transition is exploring. Both tasks are the focus of huge energy and determination.

It seems to me that what is missing is the bridge between the two. A framework that can encourage and harness those human-scale changes, and ensure that they are adequate to meeting our national commitments to reduced emissions.

Of course the UK Government (in common with others around the world) has started thinking about this challenge. Indeed, at present it has over a hundred policies that impact on emissions levels. But it has produced, in the words of its own Parliamentary Environmental Audit Committee:

> A confusing framework that cannot be said to promote effective action on climate change.
> (Environmental Audit Committee 2007)

One alternative that could provide the cohesive framework we require is David Fleming's TEQs (Tradable Energy Quotas) scheme. He first published on the scheme back in 1996, and it has been the subject of Parliamentary interest since 2004, with a scoping study (2006) and pre-feasibility study (2008) followed by a detailed report from the All Party Parliamentary Group on Peak Oil (2011). The chair of that group, John Hemming MP, has announced that:

> I believe TEQs provide the fairest and most productive way to deal with the oil crisis and to simultaneously guarantee reductions in fossil fuel use to meet climate change targets. (Fleming & Chamberlin 2011, p.5.)

In essence TEQs is an energy rationing scheme, which would guarantee that we achieve our national emissions targets while ensuring fair access to energy and supporting local initiatives like those of the Transition movement.

It would operate as the smaller-scale (national) system within a larger (global) framework for addressing climate change, while itself providing the larger framework for smaller-scale (local) energy descent plans.

This would also address the problem that within our current economic structures, reducing demand for fossil fuels locally, or even nationally, tends only to reduce the price of these fuels, and thus encourage greater consumption elsewhere.

Fleming's flash of genius when designing TEQs was the insight that we need to move away from a money-focused approach to problems that are not really about money. Our cultural belief in the omniscience of markets has led to a wide range of market-based approaches to addressing climate challenge, based on raising the price of carbon, but this approach has led to the inherently contradictory policy aims of trying to raise the carbon price while striving to keep energy prices low.

And, unsurprisingly, it has proved hard to gain popular support for increasing the cost of fossil fuels, since people rightly perceive that this increases their cost of living. As Fleming put it:

> At present, we have a policy-response shaped by sophisticated climate science, brilliant technology and pop behaviourism, based on simple assumptions about carrot-and-stick incentives. (Fleming & Chamberlin 2011, p.23.)

TEQs offer a fundamentally different approach. Rather than raising the price of carbon/energy and hoping that this reduces demand sufficiently, TEQs start from a strict quantity-based carbon budget and allow price to find its level in response to that. This restores straightforward motivation for individuals, organisations and nations. Once you guarantee people a fair entitlement to energy, in line with a declining cap, society can then collectively focus its attention on finding ways to thrive on reduced demand, and thus keeping the price of energy/carbon as low as possible. This is a simply-understood task that everyone can buy into with enthusiasm.

The real beauty of the scheme though is that it provides the large-scale framework to encourage and empower those small-scale solutions. It effectively converts the national carbon budget into a personal energy budget for everyone, with the clear recognition that this budget will be decreasing year on year. The variations in the national price of TEQs units would be of interest to all, and since lower demand means lower prices, the population would be encouraged not only to reduce their own energy use, but also to urge others to do so. Additionally, the substantial income from the auction of units to organisations would be accessible to communities to fund the building of new local infrastructure or otherwise support their energy transition.

It would be transparently in the collective interest to work together to find ingenious ways to increase low-carbon energy supplies, reduce demand and move towards the shared goal of living happily within our energy and emissions constraints, with the TEQs price providing a clear indicator of how well we are doing.

This cooperation is essential, since the transformation in infrastructures necessitated by climate change requires collaboration between the different sectors of society, united in a single scheme easily understood by all. It is a critical feature of TEQs that it encourages constructive interaction between households, businesses, local authorities, transport providers, national government, and so on. In short, the scheme is explicitly designed to stimulate common purpose in a nation.

We may sometimes be tempted to hold fossil-fuel companies and governments responsible for all our ills, but it must be recognised that even if they wished to they could not solve our energy problems without the engagement of the wider public. Our individual and community lifestyles need transformation too, and this cannot be done for us. No system can ever relieve us of our personal responsibility, and it is essential that we all recognise the need to change the way we live.

As Lord Smith of Finsbury, Chairman of the UK Environment Agency, has said:

> Rationing is the fairest and most effective way of meeting Britain's legally binding targets for cutting greenhouse gas emissions.
> (Webster 2009.)

Rationing has acquired a bad name with many due to its association with shortage, yet it is a response to shortage, not the cause of it. Combining the necessary reductions in the use of high-carbon fuel with the depletion of global energy resources is sure to put increased pressure on energy supplies, and in times of scarcity we cry out for guaranteed fair shares. The purpose of TEQs is to share out fairly the shrinking energy/carbon budget, while allowing maximum freedom of choice over energy use. The alternative is 'rationing by price' (that is, the richest get whatever is in short supply), which brings only inequity, suffering and resentment.

So while Transition takes our solitude and despair and transforms it into communal action, TEQs takes those community initiatives and combines them into an empowered – and sufficient – wave of change at the national level, ready to fulfil global agreements and resolve our global challenges.

Crucially, neither TEQs nor Transition take the top-down approach of laying out some master-plan that must then be implemented and enforced, regardless of local objections. Rather, they both provide the frameworks to support and encourage creative self-expression and cooperation. They follow the wise words of Antoine de Saint-Exupéry:

> As to the future, your task is not to foresee it, but to enable it.
> (Saint-Exupéry 1948, p.50.)

Both ideas are expressions of a very different story of our relationship with our world and with each other, and Charles Eisenstein provides a beautiful name for this story. We have been treating our Mother Earth as someone that we can take and take from without consideration for how much she can give. But Eisenstein suggests that we instead tell the story of Lover Earth:

> The relationship to a lover is different: to a lover we desire to give as well as to receive, and we desire to create together, each offering our gifts toward a purpose transcending each of us, so that our union becomes greater than the sum of our individuality.
> (Eisenstein 2009)

Let us fall madly in love with the planet with which we are profoundly interdependent, and co-create a wonderful future together. This appears to be the only story that has a future.

Of course this is no guarantee that it will come to shape our collective choices, to be told and retold. But I for one find it an inspiring and compelling vision, and I feel myself come alive when I am doing all I can do to embed it in my actions and in our culture. As Paul Wellstone once said, if we don't strive hard enough for the things we stand for, at some point we have to recognise that we don't really stand for them.

I have found the story that I want my life to tell.

References

Anderson, R. (1999) Speech to business leaders at North Carolina State University, featured in the 2004 film *The Corporation*, Big Picture Media Corporation.

Chamberlin, S. (2009) *The Transition Timeline: for a local, resilient future*, Green Books, UK.

Eisenstein, C. (2009) 'Rituals for Lover Earth,' *Dark Optimism*. At: http://www.darkoptimism.org/2009/10/16/rituals-for-lover-earth/

Environmental Audit Committee (2007) *Environmental Audit Committee Ninth Report*, House of Commons Parliamentary Press, U.K. At: http://is.gd/AXuxGr

Fleming, D. (2007) *Energy and the Common Purpose: Descending the Energy Staircase with TEQs (Tradable Energy Quotas)*, third ed., The Lean Economy Connection. At:http://www.teqs.net/downloads.html

Fleming, D. & Chamberlin, S. (2011) *TEQs (Tradable Energy Quotas): A Policy Framework for Peak Oil and Climate Change*, London: All-Party Parliamentary Group on Peak Oil, and The Lean Economy Connection. At: http://teqs.net/report/

Hansen, J. *et al.* (2008) 'Target Atmospheric CO_2: Where Should Humanity Aim?' *The Open Atmospheric Science Journal*, 2, pp.217–231. At: http://is.gd/HDVTW9

Hawken, P. (2008) *Blessed Unrest: how the largest movement in the world came into being and why no one saw it coming*, Viking, US.

Hopkins, R. (2006) 'Building Miles,' *Resurgence*, 236. At: http://is.gd/LivC6H

—, (2008) *The Transition Handbook: From oil dependency to local resilience*, Green Books, UK.

Saint-Exupéry, A. de (1948) *The Wisdom of the Sands*, Harcourt Brace & Co., NY.

Sherman, D.K. & Cohen, G.L. (2006) 'The psychology of self-defense: self-affirmation theory.' In Zanna, M.P. (ed.) *Advances in Experimental Social Psychology*, Vol. 38 (pp.183–242), San Diego, CA: Academic Press. At: http://is.gd/SsG8hM

Webster, M. (2009) 'Carbon ration account for all proposed by Environment Agency,' *The Times*. At: http://www.timesonline.co.uk/tol/news/environment/article6909046.ece

13. Resilient Economics

MARK BURTON, BRIAN GOODWIN, STEPHAN HARDING,
SERGIO MARASCHIN, AND JULIE RICHARDSON

Introduction

This chapter explores the dynamic properties of organisms and ecosystems that make them so resilient and capable of adapting to changing circumstances, allowing them to maintain an overall condition of coherence, wholeness and health while living in balance within the resources of the planet. We suggest some key principles that are required in order to facilitate the emergence of equivalently resilient and creative economies that integrate with the dynamics of earth evolution.

A primary effect of life on the dynamics of terrestrial processes is to accelerate the natural flows and cycles of energy and matter on earth. Life also has the effect of increasing the complexity of these interacting cycles so that they form a rich web of inter-related activities. These involve positive and negative feedback loops that provide both overall stability and adaptive resilience to the whole system, ensuring continuous creative evolutionary change and transformation. This we see in the 3.7 billion years of continuous evolution that has given rise to an immense diversity of species, from microbes and algae to giant redwoods and whales.

In contrast to this dynamically bounded web of creatively adaptive processes, our economic system produces continuous, unstable growth with destruction of cultural and species diversity through homogenisation of global life-styles among humans, the most recently emerged species within the Gaian complex of interacting life forms. Any species that continues to behave in this way seems bound for extinction. At this moment, when the consequences of our economic activities have become clear in the destructive instabilities generated by continuous growth and unregulated capitalism, we have a choice. We can either attempt to restore the economic system using the same basic principles as those prevailing for two hundred years, with some corrective modifications; or we can re-examine fundamental economic principles using insights from biological evolution and ecosystem dynamics to establish a radically different foundation for trade and commerce. The latter is what we explore in this chapter.

Effect of life on earth.

Life captures much of the abundant energy flux from the sun to create the extraordinary diversity of species that has emerged throughout the past three to four billion years of Earth's evolution. As shown by James Lovelock in his Gaia theory, the interaction of life with the planet's geophysical processes maintains this condition of continuous creativity and abundance on our planet as an integrated whole. It is the dynamics of this process that we need to understand in order to build the resilient wisdom and adaptability of Gaia into our economic system. How does life transform the planet from dying to living?

Life captures solar energy by slowing down the inevitable loss of heat from the Earth's surface through the creation of many cycles of production that are all coupled to each other, the output of one being the input to another. There is evidence that the main biogeochemical cycles were essentially operating at around 3.5 billion years ago and that a cycling system similar to the present day Earth has been functioning for 2.3 billion years. Life modulates biogeochemical cycles by substantially increasing rates of energy-matter transformation while increasing the overall efficiency of the system by complex feedback loops of energy-matter within the system. Some of the feedback loops act as blockers of the flow of matter or valves diverting energy to other processes (negative feedbacks), thus achieving a beautiful balance of self-regulation between flow rate and effectiveness of output. On a Gaian scale, these biogeochemical cycles co-evolved, forming a system of nested cycles that set limits on external energy required to sustain the system.

Figure 1 describes this process by comparing the effect of life on energy capture and transfer on earth with energy loss on a dead planet. On the left we see the many little coupled cycles that arise from the activities of different species, all coupled together in ecosystems so as to produce a coherent process that is the living, evolving earth. On the right is the way energy is lost from a dead planet. The same total loss of energy occurs, but on a living planet it is dramatically delayed through the sharing of the energy among the life cycles of different species.

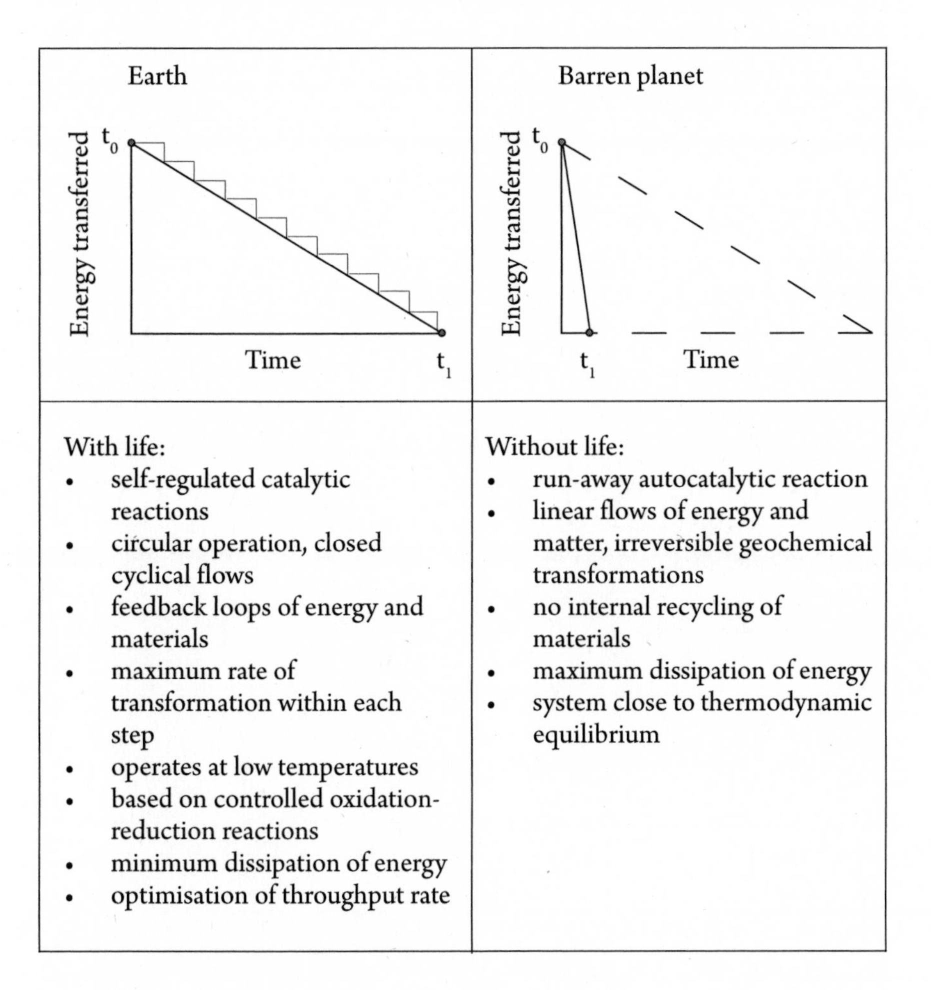

Figure 1. Patterns of energy loss from a living planet like Earth, in comparison with energy loss from a dead planet such as Mars or Venus.

Evolution brings into existence a remarkable diversity of life forms with minimal energy loss (maximum efficiency) and maximal creativity. In contrast to this, our economic system creates products using inefficient processes that accelerate the rate of energy loss and resource depletion. We see this most dramatically in our dissipation of carbon and energy by our use of fossil fuels during the past two hundred years: we have burned up vast amounts of stored carbon and energy from subterranean deposits that were deposited as part of the process whereby the Earth maintained habitable temperatures for billions of years through regulation of CO_2 in the atmosphere by means of carbon burial. This dissipation has been a runaway process, like an explosion in comparison with normal evolutionary rates of change: a very rapid, inefficient dissipation of energy that has been extremely destructive to the health of other species and the planet as a whole, as observed in the rate of species extinctions during the twentieth century. These economic activities take the Earth in the direction of a dying, not a living, planet. Our economic system generally proceeds in this manner, using up natural resources such as iron and copper very inefficiently with toxic waste products rather than useful outputs for productive activities, causing soil erosion through inefficient farming practices, and fishing with methods that cause immense destruction to fishery ecosystems in the oceans, to mention just a few examples. What are the basic principles whereby natural systems achieve their cooperative, creative efficiency, and how can we use these in our economic and technological systems? It is useful to start by looking at the way these principles work within individual organisms, and then to see how they are extended to ecosystems.

Basic postulate: catalysts facilitate flows in living systems

Living organisms are fundamentally characterised by a rich network of metabolic cycles that bring about the transformation of one type of molecule into another so that there is a continuous flux of materials through all the parts of the system. The range of metabolites is immense: several hundred thousand different molecular species are continuously synthesised and degraded in a process that maintains an appropriate balance among all the different substances within organisms. These transformations occur at room temperature. They do not require high temperatures and pressures, as many of our synthetic chemical activities do. Metabolic transformations are brought about by specialised substances produced in organisms called enzymes. These are proteins, long chains of basic molecules (amino acids) that fold up into specific shapes with the ability to bind to particular metabolites and bring about their conversion to other forms. This process is called catalysis. It accelerates a thermodynamically-permitted reaction from a rate that would be almost unmeasurable at room temperature to a significant value, often in the region of mols/sec. It is this that makes life as we know it possible.

Economic implications: money as catalyst facilitates flows in economic systems

A major feature of catalysis is that the catalyst is not used up in the process so that it can continue to facilitate the reaction for the duration of its life, which for proteins is minutes to hours. In this time one molecule of catalyst can catalyse the transformation of millions of molecules of metabolite. Money in an economic system works in a similar way. It facilitates or accelerates trade transactions that could occur by other means, such as barter; and it is not used up in these transactions so that a unit of currency can be used many thousands of times to facilitate exchange. However, there is a significant difference here from enzyme catalysis: whereas the action of any particular enzyme is specific to a small set of catalytic transformations, facilitating the conversion of one or a few metabolites into other metabolites, money can be used in any trade transaction within a global trading and exchange system. Whereas enzymes are specifically designed for particular transformations, money as we currently use it is non-specific. However, there is no reason why money should not be specifically designed to facilitate particular types of transaction rather than serving as a totally non-specific currency. We will argue that such a diversification of currencies is an essential ingredient of a resilient economic system. We shall return to this question of specificity and design of currencies, especially in connection with local community building, a crucial aspect of stable trading.

Another property of catalysts is that they do not accumulate in quantity. They have a well-defined lifetime so that they are degraded and recycled after a relatively short period of time relative to the lifetime of an organism in which they function. Whereas an organism may live for many years, a catalyst lasts for minutes to hours. Thus unused catalysts are not saved but are degraded and recycled. Similarly, money as a catalyst of exchange should not accumulate but should be continuously used to facilitate transactions. This is realised by the principle of demurrage wherein money progressively loses value if it is stored and not used. This prevents money from accumulating in anyone's hands and results in continuous, efficient facilitation of trade in the economic system.

The fact that money is seen as having value in itself so that people accumulate it arises from a confusion between goods which satisfy basic needs and money as potential for acquiring goods. The source of this confusion lies in the way we generate and distribute money, not in any intrinsic properties of money itself. While some goods will inevitably get scarce at times, in general the diversity of products available can satisfy people's needs, though not their greeds, as Gandhi pointed out. Sufficiency is the key. The monetary specialist Bernard Lietaer says the following in an interview examining the foundations of our financial system:

> I believe that greed and competition are not a result of immutable human temperament; I have come to the conclusion that greed and fear of scarcity are in fact being continuously created and amplified as a direct result of the kind of money we are using. For example, we can produce more than enough food to feed everybody, and there is definitely enough work for everybody in the world, but there is clearly not enough money to pay for it all. The scarcity is in our national currencies. In fact, the job of central banks is to create and maintain that currency scarcity. The direct consequence is that we have to fight with each other in order to survive.
> (See: http://www.transaction.net/press/interview/lietaer0497.html)

The confusion between money as facilitator of trading transactions and as something with intrinsic value comes from a failure to distinguish between *oikos,* the Greek root for 'economic', and *krema,* the Greek word for individual wealth which is purely about acquisition. Whereas *oikos* concerns the satisfaction of real needs in society (Max-Neef 1991) based on cooperative household management, modelled on nature, the 'krematistic' accumulation of money in individual hands was condemned by Aristotle as destructive of community wealth and the intrinsic health of resilient trading and exchange systems. It is the development of an economic culture based on pure moneymaking and acquisition that has shaped our monetary and economic systems, rather than ecological principles of management, so that they are intrinsically unstable and destructive.

Ecosystem catalysts

What acts like a catalyst in ecosystems, accelerating the flows of energy and matter between species? There are many candidates for this, but primary ones are bacteria. They are ubiquitous on earth and immensely diverse in their activities. These unicellular life forms use the organic debris of other species such as shed leaves, dead bodies of insects and other creatures as a source of energy and matter, producing simple molecules such as carbon dioxide, oxygen and methane as their waste products, which are released into the atmosphere. Bacteria act in ecosystems like enzymes within organisms, catalysing the transformation of waste organic material into substances that are essential for the dynamic equilibrium of the planet as a whole. Without bacteria the Earth would have no stability in temperature or gaseous composition of the atmosphere that maintains conditions suitable for life itself. This is the insight of the Gaia hypothesis that recognised the intimate connection between life and the conditions that allow for its continuous evolution on Earth (Lovelock 2000, 2005).

Bacteria achieve their extraordinary powers of transformation by using open-source methods of sharing their evolutionary discoveries with each other, and with other species. They do this by exchanging their genes, so that if one species of bacterium discovers a more efficient way of using organic matter for its metabolism by generating a new gene for an enzyme, this information is passed around to others. The bacterial world discovered the advantages of a bacterial worldwide web billions of years ago and have evolved successfully in this mode for aeons. Furthermore, different species of bacteria act in cooperatives, aggregations of individuals that take decisions collectively to adopt strategies of survival appropriate to ambient circumstances. For example if local food resources become scarcer or water supply is diminished, the colony as a whole adopts a foraging strategy that is more efficient and preserves water more effectively. This is reflected in the morphology of the colony (Ben Jacob *et al.* 2006). The ultimate strategy for surviving difficult conditions is for the colony to form spores. These are effectively forms of suspended animation in which nearly all the metabolic transformations of the organism cease, with the cell adopting a virtually solid-state or quasi-crystalline condition which is resistant to the harshest environments.

Bacteria are not alone in carrying out essential recycling of materials and energy on which evolution depends. Many different species contribute to this by eating some forms of life and producing waste that is the source of food for others. It is the extreme diversity of the tangled web of interactions that this creates that is the basis of resilience and adaptability of life forms on earth. These interacting networks of creative agents trade in real goods: the wood of the fallen tree serves as energy and material for the growth of fungi; the insect that is captured by the Venus fly trap is digested to feed the needs of the plant and allow it to grow; the calcium released from granite rocks by the growth of lichens on its surface flows down a stream, then travels by river to the ocean where it is used by unicellular organisms to grow a protective shell around its surface. This is the continuous trade in energy and goods that maintains the health and wellbeing of the Gaian system as a cooperative, resilient, evolving network of beings that form an indissoluble unity on the planet. When we allow speculation and gambling into our economic systems in the form of hedge funds and derivatives, we introduce an unregulated source of instability into the system, violating the principles that underlie resilience and adaptability in ecosystems.

Exchange and trade in ecosystems

At the level of ecosystems, time scales of change and adaptation are extended beyond the lifetimes of organisms to many generations, but the same principle of demurrage operates in relation to facilitation of trade in goods and services by transforming catalysts. Bacteria, for example, as facilitators of recycling and exchange between species, never accumulate where they are not used. Their number is always directly related to their activity. The same is true of all other members of an ecosystem and their contribution to the resilience of the whole. Organisms accumulate only where they function as facilitators of exchange, such as herds of antelope where grass is plentiful on the African savannah, using this resource to produce body mass which then feeds predators, which themselves keep the herds of antelope on the move so that they never accumulate in any one place and destroy the grasslands by overgrazing. Compare this with our un-ecological practice of keeping vast herds of beef cattle in restricted domains near water sources, without predators to move them on, so that the grasslands become degraded and the soil erodes. Excessive accumulation of anything is a serious error in resource management, where diversity, quantity directly connected to activity, and continuous recycling are basic principles of resilience. Ecological systems produce abundance because each species has the potential for exponential growth in numbers when resources are plentiful, such as bacteria accumulating in their millions under circumstances where organic material such as leaf litter is plentiful, or antelope accumulating when grass is abundant. However, predation limits numbers and populations decrease again when conditions change to scarcity for those particular species while other species thrive under the altered circumstances, resulting in a continuous change of composition of the ecosystem with changing conditions. There is never an uncontrolled increase in populations of particular species because numbers are held in check by negative feedback processes. The equivalent of these negative feedback process in human trading systems will be considered later.

Exchange and trade in ecosystems are always in terms of real goods such as light energy absorbed by leaves to produce organic food, bacteria consumed by organisms like slime moulds in forest leaf litter, organic matter and minerals consumed by worms in soil, antelope bodies consumed by prides of lions on the savannah, and so on. Ecosystems do not engage in speculative trading in future costs of food or minerals because all exchange is in terms of energy and different forms of matter. This keeps the ecosystem grounded in reality in relation to maintenance and continuity. There is of course plenty of creativity in ecosystems. The emergence of new species and innovative partnerships between species is a domain of play with future possibilities as extant organisms explore new combinations of genes within, and interactions among, themselves. This emergence of novelty is very similar to cultural creativity, being based on a process that is sensitive to both history, through genetic inheritance, and sensitive to external context through the generation of forms with appropriate adaptation to the environment. Thus do new forms of organism and communities arise in ecosystems during their resilient evolution. We could say that an adaptive and resilient culture is founded on forms of creativity that are similarly sensitive to history and to ecological context, avoiding the errors that have caused cultures to collapse as described by Diamond (2002).

Short- and long-term investment in human economic systems and in ecosystems

We are now well aware of the dangers of short-term investment and speculation cycles in human economies, as these can lead to bubbles that burst with extensive collateral damage to all economies due to global connectedness. However, there does seem to be an intrinsic tendency in human social history for episodes of creative innovation, during which new modes of living, technologies and power structures are explored in response to changing circumstances within society and in the environment, generating crises of transition that are then resolved through the adoption of new technologies and social structures appropriate to the new styles of living and production. During the past 230 years or so of the industrial revolution, such 'technological revolutions' have been described by the Venezuelan economist Carlota Perez (2002) as occurring over roughly fifty-year cycles. She describes these episodes in the following terms:

> Each technological revolution irrupts in the space shaped by the previous one and must confront old practices, criteria, habits, ideas and routines, deeply embedded in the minds and lives of the people involved as well as the general institutional framework, established to accommodate the old paradigm. This context, almost by definition, is inadequate for the new.
> (Perez 2007)

Significantly, Perez demonstrates that each transition goes through distinct periods, starting off with an 'irruption' phase during which the innovations are generated, followed by a phase of 'frenzy' as investors rush for a stake in the businesses spawned by the innovations; then a phase of 'synergy' as the new approach results in a generalised global dispersion of production systems across the economy as a new 'golden age'.

The first of these recent transitions described by Perez was effectively the beginning of the industrial age, starting around 1771 with the development of machines and the emergence of the mechanised cotton industry as the template for industrialisation. The second transition was the 'age of steam and railways' that started in 1829 when the steam engine fuelled by coal made it possible to build transportation systems and factories powered by fossil fuels extracted from the earth using new mining technologies. Subsequent cycles were the age of steel, electrification, steel ships and the start of mass consumption of consumables (starting in 1875), the age of oil, automobiles and mass production that began in 1908, then came the age of information and telecommunications starting in the USA in 1971, which spread across the world. Each of these transitions depended on the exploitation of a key natural resource/ ecosystem service: soils, cotton and iron from the colonies for the first; coal and iron ore for the second; coal, iron ore copper and agricultural produce for the third; oil plus all the other resources for the fourth; and a conglomeration of oil, metals, biomass and agricultural produce secured via digitally networked trading relationships for the fifth.

These recent transition cycles need to be embedded in the much longer transition periods that human societies and their ecosystems have undergone, though these are more difficult to characterise and much less well understood in their origins and their impacts. Among them are the discovery and use of fire by humans some 250,000 years ago, an innovation that has had serious impacts on the ecosystems within which humans have developed their societies. Another is the emergence of language, dating from some 40,000 years ago. Language clearly facilitates communication and stimulates creativity such as tool-making, and cooperative activities such as hunting and construction of dwellings, which have significant ecosystem impacts. The transition to agriculture some 10,500 years ago was another major transformation of life-style that has had lasting consequences on the health and diversity of ecosystems and species. All of these reveal periods of innovation followed by ecological impacts, the prehistoric transitions having much longer intervals between them than the recent ones described during the industrial age. As we are all well aware, human history is speeding up. However, it may well be that periods of innovation followed by episodes of consolidation are also intrinsic to ecosystem dynamics, so that human cultures are following natural dynamic patterns.

Within ecosystem evolution, clearly the short-term innovators are the microbes and viruses with their open source information sharing that allow new adaptive discoveries to be made available to other microorganisms. This can be seen from a human perspective in the rapid adaptation of bacteria to our health defence discoveries such as drugs and antibiotics. Bacteria evolve new enzymes capable of destroying antibiotics and de-toxifying drugs so that they become resistant to these products. They also learn to change their identity markers on the cell surface so that our immune systems fail to identify and destroy them, or to continually alter their genetic structure so that the immune system cannot recognise them as foreign, as in the strategy of HIV. Bacteria and viruses have behaved this way throughout evolution, adapting to changing circumstances and learning new strategies of living on their hosts, the larger animals and plants. It is the macroflora and macrophyta that are the long-term investors in ecosystems, having much longer life-cycles than microbes and hence being much slower to adapt to changing circumstances. They depend upon stability in the ecosystem, much as long-term economic investors depend upon stability in economic policy and conditions of investment. Whereas microbes can adapt to major changes in conditions on Earth, such as the emergence of oxygen as a major component of the atmosphere during the transition from an anaerobic to an aerobic planet resulting from the innovation of photosynthesis, the macrobiota respond much more slowly and need long periods of stability to discover effective life-cycle strategies. They then contribute substantially to this stability by introducing complex diversity into ecosystems, enhancing resilience for the evolutionary process.

Resilience and diversity are intimately linked and we need to take account of this in our economic systems, which have recently suffered from a combination of global homogenisation of economic systems and the destabilising effects of short-term investment activities whose sole goal is profit, not diversifying the economy with real innovations that can help to stabilise the process.

To be creatively resilient and adaptable, economies require similarly rich webs of interaction between companies of different size and complexity that trade in real goods and recycle resources efficiently among each other. Businesses clearly go through phases of growth and expansion as they discover a new niche for trading, followed by downsizing and possible extinction as their contributions to and relevance for the trading system decline.

The resilient diversity of Gaia is based on local bioregions

The diversity of interacting life forms that give Earth its resilient diversity is based on the evolutionary adaptation of organisms to local bioregions such as rainforests, savannahs, coral reefs, wetlands, and so on. Thus the foundation of global stability in Gaia is patterns of interaction in local communities that have evolved to survive in different conditions (Harding 2009). Global adaptability and resilience is grounded in appropriate local behaviour. It is not based on the application of a single principle of exchange and trade in a homogeneous system, as is our global economic system. In fact we cannot design rationally a global economic system that would ever work because this is precisely the wrong approach. Evolution experiments with a diversity of local life-cycle strategies that co-evolve to give coherent patterns of interaction. Similarly, we need to explore a diversity of economic and currency systems that are appropriate to the cultures that have emerged in different bioregions. We know that the different human languages and cultures that have evolved are deeply sensitive to their ecological contexts, reflecting an intimate understanding of the subtleties of appropriate relationship to other species, the land and its climate, the seasons and their rhythms. This sensitivity we have lost by imposing a global economic system on the planet that is based on precisely the wrong principles for resilient evolution. As a result we are losing human cultures and languages, as well as species and ecosystems, at an alarming rate. The great mistake of economic design up to now has been the assumption that we can rationally put in place a system that will work globally, once and for all. The destructive human behaviour that results from the hubris of this type of belief is very baldly expressed by Naomi Klein in her thoroughly researched and documented book *The Shock Doctrine* (2007) in connection with the doctrine of neoclassical economics, the most recent attempt to impose a single economic and trading system on the planet:

> ... the entire thirty-year history of the Chicago School experiment has been one of mass corruption and corporatist collusion between security states and large corporations, from Chile's piranhas, to Argentina's crony privatisations, to Russia's oligarchs, to Enron's energy shell game, to Iraq's 'free fraud zone'. The point of shock therapy is to open up a window for enormous profits to be made quickly – not despite the lawlessness but precisely because of it. 'Russia has become a Klondike for International Fund Speculators', ran a headline in a Russian newspaper in 1997, while Forbes described Russia and central Europe as 'the new frontier'. The colonial-era terms were entirely appropriate. This is dissipation of earth resources at a deadly rate, moving the earth towards the condition of a dying planet.

A similar type of universal vision lies behind the belief of modern Western science in finding 'the truth' and explaining the world in terms of a few basic principles that allow us to exercise control over nature. Chaos, complexity, and Gaia theories revealed the limitations of such a vision, useful as it is in restricted contexts, for nature is intrinsically and unpredictably creative. This new integrated or holistic vision requires a move to transdisciplinarity in our educational systems (Max-Neef 2005). Designing successful economies in such an interconnected world requires that we be as intrinsically and unpredictably creative as our natural context, so we must proceed by inspiration, humility, trial and error.

Exploring resilient economic systems

> *What is the meaning of democracy, freedom, human dignity, standard of living, self-realisation, fulfilment? Is it a matter of goods, or of people? Of course it is a matter of people. But people can be themselves only in small comprehensible groups. Therefore we must learn to think in terms of an articulated structure that can cope with a multiplicity of small-scale units.*
>
> (E.F. SCHUMACHER, *SMALL IS BEAUTIFUL)*

Fritz Schumacher has been an inspiration for the environmental movement and as a refuge for economists seeking a more humane and ecological approach. His insights for a re-localised economy comprising a multiplicity of small-scale units can be extended to enable sharing of information and knowledge both within and across local communities, enabled by the internet. Maximising autonomy at the micro level combined with maximising coherence at the macro level is a characteristic of health and resilience in ecosystems that has clear applications to our economic systems.

There is a basic principle that is fundamental to our belief that human communities need to be organised on the appropriate local scale in order for people to achieve creatively innovative but overall stable trading and exchange systems like those that are the foundation of ecosystem stability. This is the recognition that there is a natural regulator of human trading activity that balances quantitative satisfaction of needs with qualitative value of lived experience; that is, that brings into harmonious balance quantity of goods and services with quality of life. However, this regulator is effective only when human communities have the appropriate size or scale. The reason for this is that if communities grow too large then the social structure tends to get fragmented. As a result individuals no longer have a direct experience of living in community, that is, sharing with others through direct contact and interaction in their daily activities. They have to travel long distances for work purposes so that their home and community life is restricted and there is limited sharing of experience. This shared experience is how people monitor the quality of their lives. When this monitoring occurs then people become aware of what is happening to them in relation to their potential for living a life of meaning, getting satisfaction from their relationships and feeling their own growth towards greater fulfilment. Being aware of this process is what alerts people to the destructive aspect of scale and social fragmentation, and can result in people making deliberate choices for improved quality of life rather than increased quantities of consumer goods, the usual anodyne for loss of meaning in fragmented communities. Hence it can act as a regulator of excessive trading activity beyond a level that satisfies basic needs and provides the freedom for people to explore the meaning of their lives in relation to others. This acts like a negative feedback process in ecosystems, regulating growth and maintaining balance, but it is based on distinctly human qualities.

The issue of scale is critical to ecosystem health and equally to economic health. The globalised scale of our current trading, money and capital systems is extremely fragile and highly dependent on a ready and cheap supply of fossil fuels – particularly oil. This level of scale can neither be sustained nor is desirable. The rapid move towards re-localisation (with high priority on basic needs such as food and energy), and the emergence of diversified technologies, governance and legal structures and strengthened local communities is the transition which is urgently needed. The swing from globalisation to re-localisation is perhaps another feature of self regulating feedback characteristic of natural systems.

Another important side of this natural regulator is the positive effect of trade on human communities in providing people with basic needs and services so that their quality of life improves beyond the most basic level of survival. This is well recognised as the primary stimulus for increased production and the growth of the money supply so that quality of life can improve for all. However, the economic system that we currently experience fails to distribute money, goods and services equitably throughout society because money is not understood as a catalyst that should not accumulate in anyone's hands. Rather it is allowed to flow in directions resulting from positive feedback loops (to him that hath more shall be given; those with money invest with interest and their money supply grows exponentially, without bound). These are deeply destructive of social cohesion, causing the very fragmentation that prevents people from monitoring quality of life and correcting these unstable, runaway processes from taking over.

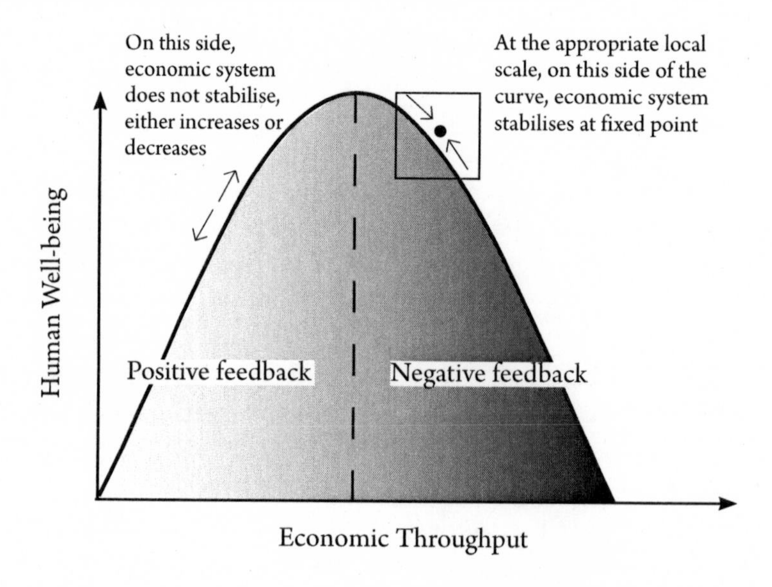

Figure 2. The dynamic regulator of human economic activity, balancing quantity of goods that satisfy basic needs with quality of life in community.

The dynamics of this regulation process is described by the graph in Figure 2. The abscissa or x-axis measures the trading activity in a community while the ordinate or y-axis defines human wellbeing in terms of the quality of life experienced. Initially, as trading activity increases in a community, facilitated by money as the medium of exchange, basic needs are satisfied and quality of life is experienced to increase as people have more opportunity to find out who they are and how their particular gifts serve their community and the broader culture. This is the part of the curve that we have been exploring through many thousands of years of human history, since the beginning of the agricultural revolution ten thousand or so years ago. The past two hundred years have seen a dramatic acceleration of the accumulation of consumer goods and money. However, at the same time there has been an enslavement of people to trading activity as work became separated from domestic and community life and people were forced to spend more and more time struggling for scarce money, an artificial aspect of our financial system.

Regulatory principles similar to those described in Figure 2 have been explored by other authors as foundations for reformed economic activities, in particular Lietaer *et al.* (2008) in a paper entitled 'White Paper on All the Options for Managing a Systemic Bank Crisis'.

Gaia is the model for the economic system that will allow humans to continue their own particular evolution on Earth in cooperation with all the other members of this planetary partnership. As recognised above, we cannot simply design a new economic system based on our limited understanding of how ecosystems work. It is necessary to approach this challenge with the creativity and uncertainty that characterises all evolutionary processes. We need to experiment with different models in different communities to find patterns of production and exchange appropriate to local regions.

What is the equivalent for human economies of the fluid, changing, and overall abundant quality of life that characterises ecosystems? As a species humans have accumulated in numbers well beyond the carrying capacity of the earth in relation to our current practices of resource management and land use. It is impossible to know exactly what level of human population can be sustained in conditions of equity and plenty when it is in balance with Earth's ecosystems, and we will only find out by exploring radically different patterns of economic practice. We suggest that some fundamental steps in this direction are the following:

1. Encourage the emergence of self-organising local communities that are based on the same principles of creative experimentation that occur in ecosystems, giving them the properties of resilient adaptation to changing circumstances. Fundamental to this process

is the recognition of the need to fundamentally change human cultural awareness from focus on gratification of human desires through consumption to satisfaction of needs and quality of life in balance with nature. This can lead to humans expressing their potential through lives of meaning realised through service to each other and to the planet. The Transition movement is a clear expression of this objective, as described in *The Transition Handbook* (Rob Hopkins 2008) and in the Structure Document for localised self-organisation (see: www.transitionculture.org.) A case study of the Transition Network is given below.

2. Encourage the localisation of food production, energy generation, healthcare, education and political decision-making so that these are all located in and around communities that have grown up within bioregions and know them well. Each community needs to explore its own form of production and community welfare, and to share with other communities the information it uses together with the results of its experiments so that an open-source network is created. This will involve the emergence of local businesses and creative innovation locally. The result is the generation of local diversity of production and social welfare systems appropriate to bioregions, regulated by local decision-making. These localisation movements can occur on different scales of regions, countries, trading networks, and so on. They will need appropriate protection legislation to allow new enterprises to develop and flourish.

    The networked environment makes possible a new modality of organising production: radically decentralised, collaborative, and non-proprietary; based on sharing resources and outputs among widely distributed, loosely connected individuals who co-operate with each other without relying on either market signals or managerial commands.
    (Yochai Benkler 2006)

3. Encourage experiments in alternative local currencies and banking practices that are based on principles of no

growth in scarce resource use and promote growth in green technologies that reduce carbon emissions. This is the Green New Deal with finance based on community banking systems that involve no interest on loans and demurrage on accumulated money that does not return to the trading system. This type of community banking already exists as evidenced by the Swedish JAK bank, time banks, and Wirtschaftsring in Switzerland. Examples of local currencies in current practice can be found in Berkshares in Vermont, the Chiemgauer currency in Germany, and in the recent introduction of the Totnes and the Lewes pounds in the UK as part of the Transition movement.

4. Stimulate the development and construction of renewable energy networks through the investment mechanism of the Green New Deal and community owned decentralised renewable energy systems. However, this involves continuous growth and so the Green New Deal needs to be embedded within a fundamentally changed economic system as described above.

   Following the radical economic thinking of Frederick Soddy (1926) who is mostly known as the 1921 Nobel Laureate in chemistry, and the economist Herman Daly (1999), the following strategic points should be considered:
   —100% reserve requirement for commercial banks, thus depriving these private institutions of the right to create and destroy money;
   —policy of maintaining a constant price index, hence keeping the purchasing power of money constant; the creation and destruction of money is vested in the authorities;
   —local currencies; freely fluctuating exchange rates.

5. We advocate an urgent monetary reform that is more aligned with ecological principles.

Putting resilience into practice: The Transition Movement

The dual challenges of peak oil and climate change have spawned a growing international network of Transition Towns (www.transitionnetwork.org). The objective of this movement is 'to support community led responses to peak oil and climate change, building resilience and happiness'. The concept of ecological resilience and its application to local economy is hard wired into the values and emerging structure of the wide network of transition communities across the globe.

Resilience replaces sustainable economic growth as the mechanism to ensure quality of life based on individual transformation and community empowerment. The Transition Network defines resilience as 'the capacity of a system to absorb disturbance and reorganise while undergoing change, so as to still retain essentially the same function, structure, identity and feedbacks' (Walker *et al.* 2004).

This is an excellent example of an experiment in self-organising local community responses to the dual shocks of climate change and peak oil. The Transition Network does not offer a standardised, homogenised plan for economic and community life beyond oil dependence. Rather it offers seven principles of transition that enable a diversified response rooted in local context. These principles are:

—Positive Visioning: Transition Initiatives are based on a dedication to the creation of tangible, clearly expressed and practical visions of community life beyond present day dependence on fossil fuels.

—Trust and Empowerment: Transition Initiatives are based on telling people the closest version of the truth that we know in times when the information available is deeply contradictory and trusting and empowering them to make appropriate responses.

—Inclusion and Openness: Successful Transition Initiatives need an unprecedented coming together of the broad diversity of society.

—Enable Sharing and Networking: Information sharing and learning are key principles of resilient ecologies that are central to transition.

—Build Resilience: How communities respond to shocks is critical to the transition path beyond fossil fuel dependency.

The movement is explicit in its intention to build resilience across key economic sectors (including food, energy and transport) and across a range of appropriate scales – from local to national.

—Inner and outer transition: Transition is a catalyst to shifting values and unleashing the energy and creativity of people to do what they are passionate about.

—Subsidiarity: Self-organisation and decision making at the appropriate scale are key principles drawn from resilient ecological systems.

Many attribute its success and phenomenal growth (over 125 initiatives spanning fourteen countries have sprung up since its initial emergence in Kinsale in 2005) to its emerging holographic structure that mimics cell growth within living organisms. The network aspires to simultaneously maximise local autonomy and maximise coherence at the macro level through shared learning and purpose.

References

Benkler, Y. (2006) *The Wealth of Networks: How Social Production Transforms Markets and Freedom,* Yale University Press, New Haven, CT.

Daly, Herman E. (1999) *Ecological Economics and the ecology of economics: essays in criticism,* Edward Elgar Publishing, Cheltenham, UK.

Diamond, J. (2002) *Collapse: How Societies Choose to Fail or to Survive,* Allen Lane, London.

Harding, S. (2009) *Animate Earth: Science, Intuition and Gaia,* second ed., Green Books, Foxhole, Dartington, UK.

Hopkins, R. (2008) *The Transition Handbook,* Green Books, Foxhole, Dartington, UK.

—, & Lipman, P. (2009) *Transition Network: Who We Are And What We Do,* Transition Network, Totnes.

Klein, N. (2007) *The Shock Doctrine,* Penguin Books, London.

Lietaer, Bernard (2001) *The Future of Money,* Century, London.

Lovelock, J. (2000) *Gaia: A New Look at Life on Earth,* Oxford University Press.

—, (2005) *Gaia and the Theory of the Living Planet,* Gaia Books.

Max-Neef, Manfred (1991) *Human Scale Development,* Zed Books, London.

—, (2005) 'Foundations of Transdisciplinarity.' *Ecological Economics* 53, 5.

Perez, C. (2002) *Technological Revolutions and Financial Capital: The Dynamics of Bubbles and Golden Ages,* Edward Elgar Publishing, Cheltenham, UK.

— (2007) 'Great surges of development and alternative forms of globalization.' *Working Papers in Technology Governance and Economic Dynamics* No. 15. The Other Canon Foundation (Norway) and Tallin University of Technology (Estonia).

Schumacher, E.F. (1973) *Small Is Beautiful; Economics as if People Mattered,* Blond and Briggs, London.

Soddy, Frederick (1926), *Wealth, Virtual Wealth and Debt (The Solution to the Economic Paradox),* third ed., Hawthorne, CA, Omni Publications, 1961, London.

Walker, B., Hollinger, C.S., Carpenter, S.R. & Kinzig, A. (2004) 'Resilience, Adaptability and Transformability in Social-ecological systems.' *Ecology and Society* 9 (2), p.5.

14. The Need for New Forms of Money and Investment in an Energy-Scarce World

RICHARD DOUTHWAITE

Until recently, whenever I made the case for stopping issuing money by lending it into circulation, I argued that new ways of issuing it were necessary to enable countries to curb their fossil energy use in response to the threat of climate change. These days, however, I've had to alter my argument because curbing fossil energy use is no longer something that countries can decide to do voluntarily. Instead, curbs are being forced on them by a restricted and increasingly highly-priced energy supply. Indeed, it is getting so much harder to find and produce coal and oil that the supply of oil cannot be increased any further and the supply of coal is increasing much more slowly than it did a few years ago. It is not going to be possible to maintain the supply of either fuel for long and with less energy becoming available, societies will not only need to change the way they put their money into circulation but also to avoid all long-term forms of debt. Shakespeare's maxim 'Neither a borrower nor a lender be' reflected the received wisdom of his energy-scarce age and it already makes perfect sense again in ours.

Let me explain the connection between energy and debt. World oil production has already peaked and the output of coal will reach its zenith soon. A few years after the peaks have been passed, the supply of both will decline quite rapidly. The supply of gas is more difficult to predict because a new source – shale gas – has come on stream recently and has turned the US from being a gas importer to an exporter. While the potential shale gas resource is large, one authority believes that most US producers are making losses and 'much of [the resource] is non-commercial even at price levels that are considerably higher than they are today'.[1] This is because the flow of gas from an expensive well runs down rapidly after its first few months.

A Canadian analyst, Paul Chefurka, expects world gas output to reach its peak in about fifteen years and coal within the next five.[2] He does not expect the supply of renewable and nuclear energy to expand fast enough to compensate and, as a result, he believes that humanity's total energy supply from all sources will begin to decline after 2025 and will increase only marginally before then. A similar estimate of the global energy supply was prepared for the 2007 Zero Carbon Britain report.[3] This showed a gentle decline in energy availability after 2010 and a more rapid fall after 2025. Similarly, the most recent forecast by the oil and gas geologist Colin Campbell, one of the founders of the Association for the Study of Peak Oil, indicates that the total amount of energy available from oil and gas production will decline slightly between now and 2020 and then begin a more rapid decline.[4]

There is a very close link between the world's energy supply and the total amount produced, the gross world product. Accordingly, whenever the world's energy supply actually does begin to decline, we must expect the world's incomes and output to decline too. These shrinking incomes will make debts progressively harder to repay, creating a reluctance both to lend and to borrow. Who would want to take on a debt when they fear they will have less income in future out of which to repay it? For a few years into the energy decline, the money supply will contract as previous years' debts are paid off more rapidly than new loans are taken out. The reduced supply will make it increasingly difficult for businesses to trade and assemble enough money to pay their employees and their taxes. Bad debts and bankruptcies will abound, the banking system will become insolvent and the money economy will show signs of breaking down.

Governments will try to head off the breakdown with the tool they are using during the current credit crunch – they will authorise their central banks to produce money out of nothing by quantitative easing (QE). So far, the QE money, which was produced debt-free and could have been distributed debt-free, has been passed to the banks at very low interest rates in the hope that they will resume lending to the real economy. But this is not happening on any scale because of the high degree of uncertainty amongst potential borrowers. There are very few areas in rich-country economies today where people can invest borrowed money and be fairly sure of being able to pay it back.

Designing a non-debt money system

A better system to get non-debt money into use in the real economy is therefore going to have to be found if normal trading is going to be possible. In designing such a system, the first question that needs to be asked is: 'Are governments or their central banks the right people to create the new money?' The value of any currency, even those backed by gold or some other commodity, is created by its users. This is because I will only agree to accept money from you if I know that someone else will accept it from me. The more people who will accept that money and the wider range of goods and services they will provide in return, the more useful and acceptable it is. If a government and its agencies accept it, that increases its value a lot.

As the users give a money its value, it follows that it should be issued to them and the money system run on their behalf. The government would be an important user but the currency should not be run entirely in its interest, even though it will naturally claim to be acting on behalf of society as a whole and thus the money's users. Past experience with government-issued currencies is not encouraging because money-creation-and-spend has always seemed politically preferable to tax-and-spend. Some spectacular inflations that have undermined a currency's usefulness have taken place as a result of lazy governments pumping unlimited amounts of money into an economy. At the very least, therefore, an independent currency authority needs to be set up to determine how much money a government should be allowed to create and spend into circulation from month to month and the commission's terms of reference should include a clause to the effect that it has to consider the interests of all the users in taking its decisions.

Regional rather than national currencies

Another design question is: 'Should the new money circulate throughout the whole national territory or would it be better to have a number of regional systems?' Different parts of every country are going to fare quite differently as energy use declines. Some will be able to use their local energy resources to maintain a level of prosperity while others will find they have few energy sources of their own and that the cost of buying their energy in from outside leaves them impoverished. If both types of region are harnessed to the same money, the poorer ones will find themselves unable to devalue to improve their exports and lower their imports. Their poverty will persist, just as it has done in Eastern Germany where the problems created by the political decision to scrap the ostmark and deny the East Germans the flexibility they needed to align their economy with the western one has left scars to this day.

If regional currencies had been in operation in Britain in the 1980s when London boomed while the North of England's economy suffered as its coal mines and most of its heavy industries were closed on the basis that they were uncompetitive, then the North-South gap which developed might have been prevented. The North of England pound could have been allowed to fall in value compared with the London one, saving many of the businesses that were forced to close. Similarly, had Ireland introduced regional currencies during the brief period it had monetary sovereignty, a Connacht punt would have created more business opportunities west of the Shannon if it had had a lower value than its Leinster counterpart.

Non-debt currencies should not therefore be planned on a national basis or, worse, a multinational one like the euro. The EU recognises 271 regions, each with a population of between 800,000 and 3 million, in its 27 member states. If all these had their own currency, the island of Ireland would have three and Britain 36, each of which could have a floating exchange rate with a common European reference currency and thus with each other. The euro could act as a reference currency for all the regional ones. It would cease to be the single currency and simply be a shared one instead.

The advantages of the regional currencies would be huge:

1. As each currency would be created by its users rather than having to be earned or borrowed in from outside, there should always be sufficient liquidity for a high level of trading to go on within that region. This would dilute the effects of monetary problems elsewhere.

2. Regional trade would be favoured because the money required for it would be easier to obtain. A strong, integrated regional economy would develop, thus building the region's resilience to shocks from outside.

3. As the amount of regional trade grew, seigniorage (the share of the new money issued to the government) would provide the regional authority with additional spending power. Ideally, this would be used for capital projects.

4. The debt levels in the region would be lower, giving it a lower cost structure, as much of the money it used would be created debt free.

User-created currencies

In addition to the regional currencies, user-created currencies need to be set up locally to provide a way for people to exchange their time, human energy, skills and other resources without having to earn their regional currency first. One of the best-known and most successful models is Ithaca Hours, a pioneering money system set up by Paul Glover in Ithaca, New York, in 1991 in response to the recession at that time. Ithaca Hours is mainly a non-debt currency since most of its paper money is given or earned into circulation but some small zero-interest business loans are also made. A committee controls the amount of money going into use. At present, new entrants pay $10 to join and have an advertisement appear in the system's directory. They are also given two one-Hour notes – each Hour is normally accepted as being equivalent to $10 – and get more Hours when they renew their membership each year as a reward for their continued support. The system has about 900 members and about 100,000 Hours in circulation, a far cry from the days when thousands of individuals and over 500 businesses participated.[5] Its decline dates from Glover's departure for Philadelphia in 2005, a move which cost the system its full-time development worker.

Hours has no mechanism for taking money out of use should the volume of trading fall, nor can it reward its most active members for helping to build the system up. It would have to track all transactions for that to be possible and that would require it to abandon its paper notes and go electronic. The result would be something very similar to the liquidity network system developed by Feasta, the Foundation for the Economics of Sustainability which is based in Ireland.[6] In a liquidity network, the currency units are transferred between accounts via the web or a mobile phone. A float is given to users according to the amount of trading they are doing and the number of trading partners they have. If the amount of trading someone is doing goes up, their float is increased. If they do less trade, part of their float is recovered. The first local liquidity networks should start operations during 2011.

New variants of another type of user-created currency, the Local Exchange and Trading System (LETS) started by Michael Linton in the Comox Valley in British Columbia in the early 1990s, are likely to be launched. Hundreds of LETS were set up around the world because of the recession at that time but unfortunately, most of the start-ups collapsed after about two years. This was because of a defect in their design: they were based on debt but, unlike the present money system, had no mechanism for controlling the amount of debt members took on or for ensuring that debts were repaid within an agreed time. Any new LETS-type systems that emerge are likely to be web based and thus better able to control the debts their members take on. If users can be prevented from being in debt for long periods, new-style LETS should not be incompatible with a shrinking national economy.

Complementary currencies have been used to good effect in times of economic turmoil in the past. Some worked so well in the US in the 1930s that Professor Russell Sprague of Harvard University advised President Roosevelt to close them down because the American monetary system was being 'democratised out of [the government's] hands'. The same thing happened to currencies spent into circulation by provincial governments in Argentina in 2001 when the peso got very scarce because a lot of money was being taken out of the country. These monies made up around 20% of the money supply at their peak and prevented a great deal of hardship but they were withdrawn in mid 2003 for two main reasons. One was pressure from the IMF, which felt that Argentina would be unable to control its money supply and hence its exchange rate and rate of inflation if the provinces continued to issue their own monies. The other more powerful reason was that the federal government felt that the currencies gave the provinces too much autonomy and might even lead to the break-up of the country.

New ways to borrow and finance

LETS and Liquidity Network currencies cannot be backed by anything since a promise to pay something specific in exchange for them implies a debt. Moreover, if promises are given, someone has to stand over them and that means that whoever does so not only has to control the currency's issue but also has to have the resources to make good the promise should that be required. In other words, the promiser would have to play the role that the banks currently perform with debt-based money. Such backed monies would not therefore spread financial power. Instead, they could lead to its concentration.

Even so, some future types of currency will be backed by promises. Some may promise to deliver real things, like kilowatt hours of electricity, just as the pound sterling and the US dollar were once backed by promises to deliver gold. Others may be bonds backed by entitlements to a share an income stream, rather than a share of profits. Both these types of money will be used for saving rather than buying and selling. People will buy them with their regional currency and either hold them until maturity if they are bonds, or sell them for regional money at whatever the exchange rate happens to be when they need to spend.

An energy-bond savings currency could work like this. Suppose a community wanted to set up an energy supply company (ESCo) to install and run a combined heat and power plant supplying its local area with hot water for central heating and electricity. The regional currency required to purchase the equipment could be raised by selling energy 'bonds' which promise to pay the bearer the price of a specific number of kWh on the day they mature. For example, someone could buy a bond worth whatever the price of 10,000 kWh was when that bond matured in five years. The money to redeem that bond would come from the payments made by people buying energy from the plant in its fifth year. The ESCo would also offer other bonds with different maturity dates and, as they were gradually redeemed, those buying power from the ESCo would, in fact, be taking ownership of the ESCo themselves.

These energy bonds will probably be issued in large denominations for sale to purchasers both inside and outside the community and will not circulate as money. However, once the ESCo is supplying power, the managing committee could turn it into a bank. It could issue notes for, say, 50 and 100 kWh which locals could use for buying and selling, secure in the knowledge that the note had real value as it could always be used to pay their energy bills. Then, once its notes had gained acceptance, the ESCo could open accounts for people so that the full range of money-moving services was available to those using the energy-backed units. An ESCo would be unlikely to do this, though, if people were happy with the way their regional currency was being run. Only if the regional unit was rapidly losing its value in energy terms would its users migrate to one which was not.

Reforming the banking system

A fundamental flaw in the current banking system is that all banks borrow short and lend long. This means that they are always technically insolvent and only the depositors' confidence, supplemented where necessary by state guarantees, ensures that they – and the financial system – remain in business. This timing imbalance contributed to the credit crunch when some banks, Northern Rock and Anglo Irish among them, found that they could not replace their short-term borrowings with new ones when the former had to be repaid. Ireland's two big banks, AIB and Bank of Ireland, cannot replace their short-term borrowings as I write and the only reason that they are still in business is that the Irish Central Bank and the European Central Bank have been providing the money so that short-term borrowings can be repaid as they fall due.

To avoid this situation in future, banks should be required to match the periods for which they lend with the periods for which they have borrowed from their depositors, none of whom should be overseas. Moreover all their lending should be for a limited period even if they could borrow depositor's money for longer. Bank loans should be purely to enable their customers to overcome temporary imbalances in their inflows and outflows. Long term funding should be handled on a different basis by new institutions such as those issuing energy bonds and by equity partnerships, which pay investors a share, not of their profits, but of their earnings stream.[7]

Conclusion

Up to now, the commercial banks allocated a society's money supply – they decided who could borrow, for what and how much. This determined what got done and thus the shape of the economy and society. In the future, that role will pass to those who supply its energy. Only this group will have, quite literally, the power to do anything. Money once bought energy. Now energy, or at least an entitlement to it, will actually be money and energy firms may become the new banks in the way I outlined. This makes it particularly important that communities develop their own energy supplies as fossil energy supplies run down, rather than relying on outside sources. If banks issuing energy-backed money do develop, they must be community owned.

As energy gets scarcer, its cost in terms of the length of time we have to work to buy a kilowatt-hour, or its equivalent, is going to increase. Consequently, communities which fail to develop their own energy supplies will find that they have to sell a greater and greater proportion of everything they produce to the outside world to pay for their power. Looked at the other way round, energy is cheaper today than it is ever likely to be again in terms of what we have to give up to get it. We must therefore ensure that, in our communities and elsewhere, the energy-intensive projects required to provide the essentials of life in an energy-scarce world are carried out now. If they are not, their real cost will go up and they may never be done.

Working examples of both backed and unbacked forms of modern regional and community monies are needed urgently. Until there is at least one example of a non-debt currency, other than gold, working well somewhere in the world, governments will cling to the hope that increasingly unstable national and multinational debt-based currencies will retain their value. Their efforts to ensure that they do will blight millions of lives, just as is happening in Greece and Ireland as I write.

Without equitable, locally- and regionally-controllable monetary alternatives to provide flexibility, the inevitable transition to a lower-energy, lower-income economy will be extraordinarily painful for thousands of ordinary communities, and millions of ordinary people. Indeed, their transitions will almost certainly come about as a result of a chaotic collapse rather than a managed descent and the levels of energy use that they are able to sustain afterwards will be greatly reduced. Their output will therefore be low and may be insufficient to allow everyone to survive. A total reconstruction of our money-issuing and financing systems is therefore a *sine qua non* if we are to escape a human, social and economic disaster as energy supplies decline.

Notes

1. 'Shale gas – Abundance or mirage? Why the Marcellus Shale will disappoint expectations,' by Arthur E. Berman, *Energy Bulletin,* October 2010. See online: http://www.energybulletin.net/stories/2010-10-28/shale-gas%E2%80%94abundance-or-mirage-why-marcellus-shale-will-disappoint-expectations.
2. http://www.paulchefurka.ca/WEAP/WEAP.html
3. *Zero Carbon Britain: an alternative energy strategy,* Centre for Alternative Technology, Machynlleth (2007).
4. http://aspoireland.files.wordpress.com/2009/12/newsletter100_200904.pdf
5. Personal communications from Paul Glover and Paul Strebel, August 2010.
6. See Graham Barnes, 'Liquidity Networks: a debt-free electronic currency system for communities' in *Fleeing Vesuvius: overcoming the risks of economic and environmental collapse,* edited by Richard Douthwaite and Gillian Fallon, Feasta, Dublin, 2010.
7. See Chris Cook, 'Equity partnerships – a better, fairer approach to developing land' in *Fleeing Vesuvius: overcoming the risks of economic and environmental collapse,* edited by Richard Douthwaite and Gillian Fallon, Feasta, Dublin (2010). The same approach can be used for non-land projects.

15. Energy Conservation is Key to Our Survival

RICHARD HEINBERG

A cursory examination of our current energy mix yields the alarming realisation that about 85 per cent of our current energy is derived from three primary sources – oil, natural gas, and coal – that are non-renewable, whose price is likely to trend higher (and perhaps very steeply higher) in the years ahead, whose energy returned from the energy spent on obtaining them (energy returned on energy invested, or EROEI) is declining, and whose environmental impacts are unacceptable. While these sources historically have had very high economic value, we cannot rely on them in the future; indeed, the longer the transition to alternative energy sources is delayed, the more difficult that transition will be unless alternatives can be identified that have superior economic and environmental characteristics.

A process for designing an energy system to meet society's future needs must start by recognising the practical limits and potentials of available energy sources. Since *primary* energy sources will be the most crucial ones for meeting those needs, it is important to identify those first, with the understanding that secondary sources will also play their roles, along with energy carriers (forms of energy that make energy from primary sources more readily useful – as electricity makes the energy from coal useful in millions of homes).

We can define a future primary energy source as one that meets, at a minimum, these make-or-break standards:

—it must be capable of providing a substantial amount of energy – perhaps a quarter of all the energy currently used nationally;
—it must have a net energy yield of 10:1 or more;
—it cannot have unacceptable environmental impacts; and
—it must be renewable.

Assuming therefore that oil, natural gas, and coal will have rapidly diminishing roles in our future energy mix, this leaves fifteen alternative energy sources with varying economic profiles and varying environmental impacts. Since even the more robust of these are currently only relatively minor contributors to our current energy mix, this means our energy future will look very different from our energy present. The only way to find out what it might look like is to continue our process of elimination.

If we regard large contributions of climate-changing greenhouse gas emissions as a non-negotiable veto on future energy sources, that effectively removes tar sands and oil shale from the discussion. Efforts to capture and sequester carbon from these substances during processing would further reduce their already-low EROEI and raise their already-high production costs, so there is no path that is both economically realistic and environmentally responsible whereby these energy sources could be scaled up to become primary ones. That leaves thirteen other candidates.

Biofuels (ethanol and biodiesel) must be excluded because of their low EROEI, and also by limits to land and water required for their production. (Remember: we are not suggesting that any energy source cannot play *some* future role; we are merely looking first for primary sources – ones that have the potential to take over even a significant portion of the current role of conventional fossil fuels.)

Energy-from-waste is not scalable; indeed, the 'resource' base is likely to diminish as society becomes more energy efficient.

That leaves ten possibilities: nuclear, hydro, wind, solar PV, concentrating solar thermal, passive solar, biomass, geothermal, wave, and tidal.

Of these, nuclear and hydro are currently producing the largest amounts of energy. Hydropower is not without problems, but in the best instances its EROEI is very high. However, its capacity for growth in the US is severely limited, and worldwide it cannot do more than triple in capacity. Nuclear power will be slow and expensive to grow. Moreover, there are near-term limits to uranium ores, and technological ways to bypass those limits (e.g. with thorium reactors) will require time-consuming and expensive research. In short, these two energy sources are unlikely candidates for rapid expansion to replace fossil fuels.

Biomass energy production is likewise limited in scalability, in this case by available land and water, and by the low efficiency of photosynthesis. America and the world could obtain more energy from biomass, and biochar production raises the possibility of a synergistic process that would yield energy while building topsoil and capturing atmospheric carbon (though some analysts doubt this because pyrolysis emits not only CO^2 but other hazardous pollutants as well). In any case, much more research is needed before biochar production is ready for industrial-scale deployment. Competition with other uses of biomass for food and for low-energy input agriculture will also limit the amount of plant material available for energy production. Realistically, given the limits mentioned, biomass cannot be expected to sustainably produce energy on the scale of oil, gas, or coal.

Passive solar is excellent for space heating, but does not generate energy that could be used to run transportation systems and other essential elements of an industrial society.

That leaves six sources: Wind, solar PV, concentrating solar thermal, geothermal, wave, and tidal – which together currently produce only a tiny fraction of total world energy. And each of these still has its own challenges – like intermittency or limited growth potential.

Tidal, wave power, and geothermal electricity generation are unlikely to be scalable; although geothermal heat pumps can be used almost anywhere, they cannot produce primary power for transport or electricity grids.

Solar photovoltaic power is still expensive. While cheaper PV materials are now beginning to reach the market, these generally rely on rare substances whose depletion could limit deployment of the technology. Concentrating PV promises to solve some of these difficulties; however, more research is needed and the problem of intermittency remains.

With good geographical placement, wind and concentrating solar thermal have good net energy characteristics and are already capable of producing power at affordable prices. These may be the best candidates for non-fossil primary energy sources – yet again they suffer from intermittency.

Thus there is no single 'silver-bullet' energy source capable of replacing conventional fossil fuels directly – at least until the problem of intermittency can be overcome – though several of the sources discussed already serve, or are capable of serving, as secondary energy sources.

This means that as fossil fuels deplete, and as society reduces reliance on them in order to avert catastrophic climate impacts, we will have to use every available alternative energy source strategically. Instead of a silver bullet, we have in our arsenal only BBs, each with a unique profile of strengths and weaknesses that must be taken into account.

But since these alternative energy sources are so diverse, and our ways of using energy are also diverse, we will have to find ways to connect source, delivery, storage, and consumption into a coherent system by way of common energy carriers.

A common carrier

While society uses oil and gas in more or less natural states (in the case of oil, we refine it into gasoline or distil it into diesel before putting it into our fuel tanks), we are accustomed to transforming other forms of energy (such as coal, hydro, and nuclear) into electricity – which is energy in a form that is easy and convenient to use, transportable by wires, and that operates motors and a host of other devices with great efficiency.

With a wider diversity of sources entering the overall energy system, the choice of an energy carrier, and its further integration with transportation and space heating (which currently primarily rely on fossil fuels directly), become significant issues.

For the past decade or so there has been some dispute as to whether the best energy carrier for a post-fossil fuel energy regime would be electricity or hydrogen. The argument for hydrogen runs as follows: Our current transportation system (comprised of cars, trucks, ships, and aircraft) uses liquid fuels almost exclusively. A transition to electrification would take time, retooling, and investment, and would face difficulties with electricity storage (discussed in more detail below): physical limits to the energy density by weight of electric batteries would mean that ships, large trucks, and aircraft could probably never be electrified in large numbers. The problem is so basic that it would remain even if batteries were substantially improved. Hydrogen could more effectively be stored in some situations, and thus might seem to be a better choice as a transport energy carrier. Moreover, hydrogen could be generated and stored at home for heating and electricity generation, as well as for fueling the family car.

However, because hydrogen has a very low energy density per unit of volume, storage is a problem in this case as well: hydrogen-powered airplanes would need enormous tanks representing a substantial proportion of the size of the aircraft. Moreover, several technological hurdles must be overcome before fuel cells – which would be the ideal means to convert the energy of hydrogen into usable electricity – can be widely affordable. And since conversion of energy is never 100 per cent efficient, converting energy from electricity (from solar or wind, for example) to hydrogen for storage before converting it back to electricity for final use will inevitably entail inefficiencies.

The problems with hydrogen are so substantial that many analysts have by now concluded that its role in future energy systems will be limited (we are likely never to see a 'hydrogen economy'), though for some applications it may indeed make sense.

Industrial societies already have an infrastructure for the delivery of electricity. Moreover, electricity enjoys some inherent advantages over fossil fuels: it can be converted into mechanical work at much higher efficiencies than can gasoline burned in internal combustion engines, and it can be transported long distances much easier than oil (this is why high-speed trains in Europe and Japan run on electricity rather than diesel).

But if electricity is chosen as a systemic energy carrier, the primary problems with further electrifying transport using renewable energy sources such as wind, solar, geothermal, and tidal power remain: how to overcome the low energy density of electric batteries, and how to efficiently move electricity from remote places of production to distant population centres.[1]

Energy storage and transmission

The energy densities by weight of oil (42 megajoules per kilogram), natural gas (55 MJ/kg), and coal (20 to 35 MJ/kg) are far higher than those of any electricity storage medium currently available. For example, a typical lead-acid battery can store about 0.1 MJ/kg, about one-fifth of one per cent of the energy-per-pound of natural gas. Potential improvements to lead-acid batteries are limited by chemistry and thermodynamics, with an upper bound of less than 0.7 MJ/kg.

Lithium-ion batteries have improved upon the energy density of lead-acid batteries by a factor of about 6, achieving around 0.5 MJ/kg; but their theoretical energy density limit is roughly about 2 MJ/kg, or perhaps 3 MJ/kg if research on the substitution of silicon for carbon in the anodes is realised in a practical way.

It is possible that other elements could achieve higher energy storage by weight. In principle, compounds of hydrogen-scandium, if they could be made into a battery, could achieve a theoretical limit of about 5 MJ/kg. Thus the best existing batteries get about 10 per cent of what is physically possible and 25 per cent of the demonstrated upper bound.

Energy can be stored in electric fields (via capacitors) or magnetic fields (with superconductors). While the best capacitors today store one-twentieth the energy of an equal mass of lithium-ion batteries, a new company in Texas called EEstor claims a ceramic capacitor capable of 1 MJ/kg. Existing magnetic energy storage systems store around 0.01 MJ/kg, about equal to existing capacitors, though electromagnets made of high-temperature superconductors could in theory store about 4 MJ per litre, which is similar to the performance of the best imaginable batteries.

Chemical potential energy can be stored as fuel that is oxidised by atmospheric oxygen. Zinc air batteries, which involve the oxidation of zinc metal to zinc hydroxide, could achieve about 1.3 MJ/kg, but zinc oxide could theoretically beat the best imagined batteries at about 5.3 MJ/kg.

Once again, hydrogen can be used for storage. Research is moving forward on building-scale systems that will use solar cells to split water into hydrogen and oxygen by day and use a fuel cell to convert the gases to electricity at night.[2] However, as discussed above, this technology is not yet economical.[3]

Better storage of electricity will be needed at several points within the overall energy system if fossil fuels are to be eliminated from it: not only will vehicles need efficient batteries, but grid operators relying increasingly on intermittent sources like wind and solar will need ways to store excess electricity at moments of over-abundance for times of peak usage or scarcity. Energy storage on a large scale is already accomplished at hydroelectric dams by pumping water uphill into reservoirs at night when there is a surplus of electricity: energy is lost in the process, but a net economic benefit is realised in any case. This practice could be expanded, but it is limited by the number and size of existing dams, pumps, and reservoirs. Large-scale energy storage by way of giant flywheels is being studied, but such devices are likely to be costly.

The situation with transmission is also daunting. If large amounts of wind and solar energy are to be sourced from relatively remote areas and integrated into national and global grid systems, new high-capacity transmission lines will be needed, along with robust two-way communications, advanced sensors, and distributed computers to improve the efficiency, reliability, and safety of power delivery and use.

For the US alone, the cost of such a grid upgrade would be $100 billion at a minimum, according to one recent study.[4] The proposed new system that was the basis of the study would include 15,000 circuit miles of extremely high voltage lines, laid alongside the existing electric grid infrastructure, starting in the Great Plains and Midwest (where the bulk of the nation's wind resources are located) and terminating in the major cities of the East Coast. The cost of building wind turbines to generate the amount of power assumed in the study would add another $720 billion, spent over a fifteen-year period and financed primarily by utilities and investors. However, this hypothetical project would enable the nation to obtain only 20 per cent of its electricity from wind by 2024. If a more rapid and complete transition away from fossil fuels is needed or desired, the costs would presumably be much higher.

However, many energy analysts insist that long high-capacity power lines would *not* be needed for a renewable energy grid system: such a system would best take advantage of regional sources: off-shore wind in the US Northeast, active solar thermal in the desert Southwest, hydropower in the Northwest, and biomass in the forested Southeast. Such a decentralised or 'distributed' system would dispense not only with the need for costly high-capacity power line construction but would also avoid fractional power losses associated with long-distance transmission.[5] Yet problems remain: one of the advantages of a continent-scale grid system for renewables would be its ability to compensate for the intermittency of energy sources like wind and solar. If skies are overcast in one region, it is likely that the sun will still be shining or winds blowing elsewhere on the continent. Without a long-distance transmission system, there must be some local solution to the conundrum of electricity storage.

Transition plans

There is an existing literature of plans for transitioning US or world energy systems away from fossil fuels. It would be impossible to discuss those plans here in any detail. They include ones that involve nuclear power,[6] as well as those that exclude it.[7] Some see a relatively easy transition to solar and wind,[8] while others do not.[9]

A study of eighteen energy sources examined in terms of ten criteria, undertaken in 2009 by Post Carbon Institute, confirms that the transition is inevitable and necessary (due to the fact that fossil fuels are depleting and have declining EROEI) and that the transition will be neither easy nor cheap.[10] Further, it is reasonable to conclude that a full replacement of energy currently derived from fossil fuels with energy from alternative sources is probably impossible over the short term, and may be unrealistic to expect even over longer time frames. The daunting core problem consists of replacing a concentrated store of solar energy with a flux of solar energy (in any of the various forms in which it is available, including sunlight, wind, biomass, and flowing water).

Inevitably decisions about how much of a hypothetical energy mix should come from each of the potential sources (wind, solar, geothermal, and so on) depend on projections regarding technological developments and economic trends. All too often real-world political and economic events turn hypothetical energy transition scenarios into forgotten pipedreams.

The actual usefulness of energy transition plans is more to show what is possible than to forecast events. For this purpose even very simple exercises can sometimes be helpful in pointing out problems of scale. For example, the following three scenarios for world energy, which assume only a single alternative energy source using extremely optimistic assumptions, put humanity's future energy needs into a simple but helpful cost perspective.[11]

Scenario 1: The World at American Standards. If the world's population were to stabilise at nine billion by 2050, bringing the entire world up to US energy consumption (100 quadrillion Btu annually) would require 6000 quads per year. This is more than twelve times current total world energy production. If we assume that the cost of solar panels can be brought down to 50 cents per watt installed (one tenth the current cost and less than the current cost of coal), an investment of $500 trillion would be required for the transition, not counting grid construction and other ancillary costs – an almost unimaginably large sum. This scenario is therefore extremely unlikely to be realised.

Scenario 2: The World at European Standards. Since Europeans already live quite well using only half as much energy as Americans do, it is evident that a US standard of living is an unnecessarily high goal for the world as a whole. Suppose we aim for a global per-capita consumption rate 70 per cent lower than that in the United States. Achieving this standard, again assuming a population of nine billion, would require total energy production of 1800 quads per year, still over three times today's level. Cheap solar panels to provide this much energy would cost $150 trillion, a number over double the current world annual GDP. This scenario is conceivable, but still highly unlikely.

Scenario 3: Current per-Capita Energy Usage. Assume now that current world energy usage is maintained on a per-capita basis (if people in less-industrialised nations are to consume more, this must be compensated for by reduced consumption in industrial nations), again with the world's population stabilising at nine billion. In this case the world would consume 700 quads of energy per year. This level of energy usage, if it were all to come from cheap solar panels, would require $60 trillion in investment – still an enormous figure, though one that might be achievable over time. (Current average per-capita consumption globally is 61 gigajoules per year; in Qatar it is 899 GJ per year, in the US it is 325 GJ per year, in Switzerland it is 156 GJ per year, and in Bangladesh it is 6.8 GJ per year. The range is very wide. If Americans were to reduce their energy use to the world average, this would require a contraction to less than one-fifth of current consumption levels, but this same standard would enable citizens of Bangladesh to increase their per-capita energy consumption nine-fold.)

Of course, as noted above, all three scenarios are extremely simplistic: on one hand, they do not take into account amounts of energy already coming from hydro, biomass, and so on, which could presumably be maintained: it would not be necessary to produce all needed energy from new sources. But on the other hand, costs for grid construction and electrification of transport are not included. Nor are material resource needs accounted for. Thus on balance, the costs cited in the three scenarios are if anything probably substantially understated.

The conclusion from these scenarios seems inescapable: unless energy prices drop in an unprecedented and unforeseeable manner, the world's economy is likely to become increasingly energy-constrained as fossil fuels deplete and are phased out for environmental reasons. It is highly unlikely that the entire world will ever reach an American or even a European level of energy consumption, and even the maintenance of current energy consumption levels will require a massive level of investment.

The case for conservation

The problem of how to continue supplying energy in a world where resources are limited becomes much easier to solve if we find ways to proactively reduce energy demand. And that project in turn becomes easier if there are fewer of us wanting to use energy (that is, if population shrinks rather than continuing to increase).

The conclusion that we will probably have less energy to use, though not yet included in official projections from the International Energy Agency, seems well supported by the analysis here. Fossil fuel supplies will probably decline faster than alternatives can be developed to replace them. New sources of energy will in many cases have lower EROEI profiles than conventional fossil fuels have historically had, and will require expensive new infrastructure to overcome problems of intermittency. Moreover, the current trends for energy demand reduction and for falling investment in new energy supplies (especially from fossil fuels, but other energy sources as well) resulting from the ongoing global economic crisis, are likely to continue for several years.

How far will supplies fall, and how fast? Taking into account depletion-led declines in oil and natural gas production, a levelling off of energy from coal, and the recent shrinkage of investment in the energy sector, it may be reasonable to expect a reduction in global energy availability of 25 per cent or more during the course of the next twenty-five years. Factoring in expected population growth, this implies substantial per-capita reductions in available energy. These declines are unlikely to be evenly distributed among nations, with oil and gas importers being hardest hit.

Thus the question the world faces is not *whether* to reduce energy consumption, but *how*. Policy makers could choose to manage energy unintelligently (maintaining fossil fuel dependency as long as possible while making both poor choices of alternatives and insufficient investments in them), in which case results will be catastrophic. Transport systems will wither, global trade will contract dramatically, and energy-dependent food systems will falter, leading to very high long-term unemployment and perhaps even famine.

However, if policy makers manage the energy downturn intelligently, an acceptable quality of life could be maintained in both industrialised and less-industrialised nations; at the same time, greenhouse gas emissions could be reduced dramatically. This would require:

—re-localisation of much economic activity (especially the production and distribution of essential bulky items and materials) in order to lessen the need for transport energy;[12]
—construction of highly efficient rail-based transit systems and the redesign of cities to reduce the need for human transport;[13]
—retrofit of building stock for maximum energy efficiency (energy demand for space heating can be dramatically reduced through super-insulation of structures and by designing to maximise solar gain);[14]
—redesign of food systems to substantially reduce energy inputs;[15] and
—reduction of the need for energy in water pumping and processing through intensive water conservation programs (considerable energy is currently used in moving water, which is essential to both agriculture and human health).[16]

The goal of all these efforts must be the realisation of a steady-state economy, rather than a growth-based economy. This is because energy and economic activity are closely tied: without growth in available energy, economies cannot expand. It is true that improvements in efficiency, the introduction of new technologies, and the shifting of emphasis from basic production to provision of services can enable some economic growth to occur without an increase in energy consumption, but such trends have inherent bounds. Over the long run, static or falling energy supplies must be reflected in economic stasis or contraction. However, with proper planning, there is no reason why, under such circumstances, an acceptable quality of life could not be maintained.[17] For the world as a whole, this might entail partial redistribution of energy consumption, with industrial nations reducing consumption substantially, and less-industrial nations increasing their consumption somewhat in order to foster global equity.

Notes

1. 'Some Thoughts on the Obama Energy Agenda from the Perspective of Net Energy,' *The Oil Drum* (2009), at: http://netenergy.theoildrum.com/node/5073#more

2. Anne Trafton, '"Major discovery" from MIT primed to unleash solar revolution,' MIT News, Massachusetts Institute of Technology (2008), at: http://web.mit.edu/newsoffice/2008/oxygen-0731.html

3. Kurt Zenz House and Alex Johnson, 'The Limits of Energy Storage Technology,' *Bulletin of the Atomic Scientists* (January 2009). At: http://thebulletin.org/web-edition/columnists/kurt-zenz-house/the-limits-of-energy-storage-technology

4. Rebecca Smith, 'New Grid for Renewable Energy Could Be Costly,' *The Wall Street Journal* (February 9, 2009). At: http://online.wsj.com/article/SB123414242155761829.html

5. Ian Bowles, 'Home-Grown Power,' *New York Times*, March 6, 2009. At: www.nytimes.com/2009/03/07/opinion/07bowles.html

6. Intergovernmental Panel on Climate Change, *Climate Change 2007 – Mitigation of Climate Change*, Working Group III, Fourth Assessment Report. At: http://www.mnp.nl/ipcc/pages_media/AR4-chapters.html

7. Arjun Makhijani, 'Carbon-Free and Nuclear-Free: A Roadmap for U.S. Energy Policy,' *Science for Democratic Action*, Institute for Energy and Environmental Research (August 2007). At: www.ieer.org/sdafiles/15-1.pdf

8. Rocky Mountain Institute, at: http://ert.rmi.org/research

9. Ted Trainer, *Renewable Energy Can Not Sustain a Consumer Society*, (Dordrecht NL: Springer, 2007).

10. Richard Heinberg, *Searching for a Miracle: Net Energy Limits & the Fate of Industrial Societies* (International Forum on Globalization and Post Carbon Institute, 2009). See: www.postcarbon.org/report/44377-searching-for-a-miracle.

11. Praveen Ghanta, 'How Much Energy Do We Need?', post to *True Cost*, February 19, 2009, see: http://truecost.wordpress.com/2009/02/19/how-much-energy-do-we-need/

12. Jason Bradford, 'Relocalization: A Strategic Response to Climate Change and Peak Oil,' post to *The Oil Drum*, June 6, 2007, at: www.theoildrum.com/node/2598

13. Richard Gilbert, *Transport Revolutions: Moving People and Freight Without Oil* (Earthscan, 2008).

14. Passive House Institute, 'Definition of Passive Houses,' at: www.passivhaustagung.de/Passive_House_E/passivehouse_definition.html

15. Richard Heinberg and Michael Bomford, *The Food and Agriculture Transition.* Sebastopol, CA: Post-Carbon Institute and The Soil Association, 2009.

16. Ben Block, 'Water Efficiency Key to Saving Energy, Expert Says,' *Worldwatch Institute* website, February 11, 2009, at: www.worldwatch.org/node/6007. See also California Energy Commission, *California's Water-Energy Relationship,* CEC-700-2005-011-SF (November 2005), at: www.energy.ca.gov/2005publications/CEC-700-2005-011/CEC-700-2005-011-SF.PDF

17. For a simulation of how this could work, see Peter Victor, *Managing without Growth: Slower by Design, Not Disaster,* Edward Elgar Publications, London, 2008.

16. Don't Blame the Hammer When You Hit Your Thumb! Or: How is Business Saving the World?

NIGEL TOPPING

Isn't this a perverse question in a book about problems created by business?

Surely it's business that got us into this mess! Well, yes, business is the institution which delivers the vast majority of environmental and social degradation on behalf of the world's consumers. But these businesses were built by us to serve us, working in markets constructed according to our rules. We make daily decisions which contribute to the perpetuation of this system and it is simply too easy to stand aside and point the finger of blame. The reality is much more complex and denying this will not help us make practical interventions to bring us on to a path which heals. Pointing the finger at business is about as useful as shouting at the hammer when you hit yourself on the thumb – it may provide a momentary focus for your anger and pain but it will not change the behaviour of the hammer!

This is a serious point. Many who profess a 'holistic' worldview and who delight in their deep sense of connection to the natural world and to indigenous peoples, suffer from what I call 'selective connectivity'. 'Holistic' relates to the whole – if everything is connected then we must acknowledge that we are connected to the factories and the banks just as we are connected to the flowers and the trees. Creating a false duality based on what we are comfortable with is no way to create solutions which must exist in the real, messy world.

The importance of including business in the solution

Business is the most powerful institution in the world today. It has vast resources at its disposal; intellectual, financial, human and physical. How can we conceive of a solution to global problems without thinking through the substantive role of this most important of actors? Of course there is a role for those seeking to change the system by creating alternatives to current organisational forms. Yet to imagine a solution where existing businesses cease to exist rather than mutate to thrive in a rapidly changing context would be folly. This would amount to imagining a form of systemic change only seen at the time of mass-extinctions. Some environmentalists appear to relish such a catastrophic breakdown, forgetting that the human and environmental costs would almost certainly be orders of magnitude greater than any form of evolutionary transition.

How is the business community responding?

We are collectively living well beyond our means, currently consuming the equivalent of 1.5 planet's worth of sustainable resources (Global Footprint Network 2011). Most of this is driven by large businesses around the world. And so at one level the task ahead for business is simple – deliver the goods and services needed by a global population whilst radically improving resource efficiency. A fourfold improvement should do it (Weizsacker *et al.* 1998), allowing for growth in population to around nine billion by mid-century and an increase in material consumption by the world's poorest, who surely need it. This would equate to doubling the amount of value delivered to the world whilst using half of the physical resources, getting us back to 0.75 planet's worth of sustainable resource use, with a comfortable margin of error. There will be no real choice here, just a matter of timing. As David Blood of Generation Investment Management said, following the recent economic crisis: 'Nature doesn't do bail-outs'. (Blood 2009) The finite nature of our planet means that our system of production and consumption will have to change. We have a collective choice. We could keep on running hard and fast beyond our means, willing a solution marked by catastrophe and conflict. Or we could delve deep into our collective source of creativity and transform the current system into one which works. Both are possible; the latter preferable, a mix of the two most likely.

Such transformative changes have typified the global system of capitalism since its inception. Anatole Kaletsky argues for just such a 'system upgrade' in his book *Capitalism 4.0. Most* businessmen and capitalists since Adam Smith wrote *The Wealth of Nations* in 1776 have not been able to conceive of a system constrained by lack of resources. For Smith writing at the time when much of North America was still tree-covered wilderness, this was excusable but for anyone writing in the twenty-first century, it suggests a lazy lack of research. If these changes are inevitable, businesses which innovate at the right time will gain a big advantage and thrive as wave after wave of disruptive change breaks over the old edifices of capitalism.

In order to contribute to solutions and hence to thrive, businesses are developing new capabilities in three areas: holistic leadership, innovating like nature and leading beyond the firm. Below I explore each of these areas, with illustrations to show what is already happening. I am unapologetic in taking all examples from large global companies. Whilst it is true that much innovation in any system happens at the fringes (e.g. small companies in a global economy), we need change at a massive scale in order to address issues such as climate change fast. Hence my interest in what large organisations are doing.

Developing holistic leadership

In the twentieth century businesses moved from a high degree of vertical integration (for instance, Ford mining iron ore) to a high degree of specialisation. In the manufacturing industries most firms now focus on design and assembly, perhaps retaining a small number of core technological capabilities in-house. Think about the automotive or computer industries. This has allowed management teams to focus on a small number of 'core competencies' whilst relying on the competitive nature of global markets to keep their supply chains working effectively. This was an era of gradual, linear change. When a supplier in one country became uncompetitive, it was simple enough to find and develop a supplier in a lower-cost economy. In an interconnected, resource constrained world, however, change can become non-linear.

For example, in 2004 I was responsible for global procurement for an international manufacturer of brake pads for the automotive industry. World steel prices spiked, largely as a result of Chinese consumption and the mothballing of European capacity. This affected my firm and all of our competitors equally – we had to struggle to increase prices with the world's car manufacturers, contrary to a thirty year trend of downward prices – no mean feat! But our real challenge came with a much less obvious commodity – molybdenum. We used small amounts of this specialist material to fine-tune the complex performance requirements of a brake. As we struggled to manage rising steel costs we were dismayed to find that molybdenum costs had not just doubled but gone up by a factor of 10 times. Why was this? Molybdenum is used in the manufacture of steel; hence global demand was being sucked into steel production, leaving us with a major cost problem. Our only solution was to reformulate and retest hundreds of brake pads at great expense in order to substitute antimony, a similar material. Our lack of awareness of resource constraints had left us completely unprepared for such change. No-one had ever asked the right questions.

This is just one concrete example of why companies need to develop holistic leadership capabilities as the combination of globalisation and resource-constraint change the rules of business forever.

There are four areas of holistic leadership which are emerging; learning about a resource constrained world, thinking in terms of the whole life-cycle impact of products and services, incorporating environmental and social issues into mainstream strategy and management, building long-term value.

1. Learning about resource constraints

No matter what industry a business operates in, the increasing number of resource constraints, from climate change to fish stocks to molybdenum availability, means that every sector will be exposed to non-linear risks. Simply taking the time out to educate the business will lead to changing behaviour. In December 2008 I contributed to a seminar on sustainability at Acer computers in Taiwan. Chairman J.T. Wang gathered together board members and key suppliers to explain why he thought this was an important theme. J.T. Wang has grown Acer to become one of the biggest computer manufacturers in the world. He is clearly an intelligent and very successful businessman. He explained that he had looked at sustainability a couple of years earlier and had decided that it was not relevant to Acer, that it was little more than the ramblings of a small group of environmentalists. But he was busy building his business and not able to go into the issues very deeply. By 2008 he had appointed a CEO and was able to take more time to read and reflect on changes in the world. And so he came to the conclusion that resource constraints are real and will have profound impacts on his sector and his business. Having realised this, JT could stand up in front of all his major suppliers and deliver a very clear message – either start taking sustainability seriously or you will not be a supplier of ours in the near future. Lesson number one – get informed about the scientific and physical reality of resource constraint. The evidence is all there, clearly laid out, you just have to take the time to get informed and the implications on your business will start to become clear.

2. Managing the full value-chain

I started my career as an engineer in the British manufacturing conglomerate Lucas Industries. In the late 1980s Lucas realised that the majority of its costs were no longer in its factories but with its suppliers, a result of reducing levels of vertical integration. We realised that purchasing managers were responsible for 60–90% of the costs of production, depending on the product we were manufacturing. We used to joke that people who weren't good enough to run the factories worked in quality and that people who weren't good enough to work in quality worked in purchasing! So in a sense we had our third-rate managers responsible for the majority of the business, hardly a recipe for success.

This same weakness is still sadly often true in many businesses, where the high costs and increasing complexity of supply chains are not well managed. Why is this important from a sustainability point of view? Simply put, many of the biggest environmental and social impacts of a business may well be outside its own boundaries, with suppliers or where the product is used. This is well illustrated by WalMart's work on climate change. When the Bentonville, Arkansas-based global retailer began measuring its carbon footprint, it started with its stores and transport fleet, both directly under its control. But it soon realised that all of these emissions were only a small part of a much larger whole. In fact, later work revealed that WalMart's own emissions are only approximately 8% of the total emissions associated with the production, transport, retail, use and disposal of everything it sells. The implications are important – if a company wants to reduce the impact of its activities it must understand what they are in the context of the full life cycle of products or services.

WalMart has done a lot to reduce its own emissions but it is now focusing on its extended value chain, committing to reduce emissions by 20 million tons over a five year period (WalMart 2010). Understanding the full value chain is necessary in order to know where the areas of biggest impact are. This in turn helps businesses to drive the biggest environmental returns on their investments.

3. *Incorporating ESG into mainstream management and strategy*
Within the business and investment community, non-financial issues fall under the catch-all heading of ESG (environmental, social and governance). As they have become more important to consumers, regulators and investors, companies have typically created sub-committees or appointed specialists to CSR (Corporate Social Responsibility) or Sustainability Management positions. Such managers rarely had any experience of running businesses and thus lacked the language and credibility to influence real change within a business. It is a sign of the growing importance of such issues that a new breed of manager is emerging – the Chief Sustainability Officer. The job title may vary but several characteristics are clear. These managers are highly experienced, often having led a major part of the business. They thus know the realities of the business very well, allowing them to formulate credible sustainability strategies and to use their existing network of colleagues to make real change happen. As an example of this trend, an old friend of mine has recently been appointed to the board of a US based industrial conglomerate. He has worked for them on three continents, most recently heading a division in Asia. He now sits on the board, reports directly to the CEO and has responsibility for all major cross-cutting themes, including supply chain, environment and climate change.

Further evidence of this trend towards including sustainability in mainstream management and strategy is the increasing requirement for companies to disclose environmental and social information to investors who are starting to understand the need for change. On top of the growing number of investable indices with an environmental theme and the move towards more sustainable listing requirements for quoted companies (the 'King 3' requirements in South Africa for example – Sustainability 2009), the International Integrated Reporting Committee (IIRC) has just been formed. The IIRC aims to make broad ESG disclosure mandatory around the world and has assembled an impressive cast of leaders to support this move with the adoption of an Integrated Reporting Framework by the G20 in 2011 as its goal (IIRC 2010).

4. Managing for the long term

Society needs businesses that are sustainable in every sense. Not only must they deliver products and services that people need with massively increased resource efficiency, they must employ people and remain economically viable over long periods. The recent systemic focus on short-term corporate profit has driven the depletion of social as well as natural capital. In the short-term it is easy to destroy value – cut jobs, cut training, move to low-cost countries, ignore environmental externalities that have not been regulated yet. It takes a long-term view to deliver long-term sustainable value – socially and environmentally as well as economically. However, managers are systemically pushed into short term thinking. The standard management excuse goes something like this: 'Believe me, I would love to manage for the long term but I have to deliver quarterly earnings updates to Wall Street and if I don't keep on delivering they will just get rid of me and replace me with someone who does what they want'. Whenever we hear such language we should be reminded that even those who seem to us to hold the most power in the system of capitalism often feel themselves powerless to act independently and to deviate from behavioural norms. A couple of examples of both investor and CEO behaviour recently demonstrate that another way is possible. Consider the words of the CEO of Unilever, one of the world's largest producers of consumer goods.

Interviewed in the *Financial Times,* CEO Paul Polman noted they have stopped giving earnings guidance for the next quarter. And he added that: 'we certainly don't want to attract the investor base that wants higher and higher and quicker results against targets that we put out every 90 days'. (Polman 2010)

In other words, Polman courageously tells short-term investors that he is not interested in their money!

Some investors are also starting to incorporate long-term sustainability research into their selection of companies who they believe are already doing a good job of managing long-term issues. For example, here is what Al Gore and David Blood's firm Generation Investment Management have to say about sustainability and long-term value creation:

> Sustainability issues can impact a company's ability to generate returns and therefore must be fully integrated with fundamental equity analysis for superior long term investment results. Our research focuses on long term economic, environmental, social and governance risks and opportunities that can materially impact a company's ability to sustain profitability and deliver returns.
> (Generation 2010)

Innovating like nature to create new solutions

Retaining the ability to innovate has always been a challenge for large organisations as they seek to standardise processes to replicate early successes. This challenge is amplified in the massively interconnected, heavily populated and resource-constrained world of the twenty-first century ('Hot, flat and crowded' as the title of Thomas Friedman's 2008 book on these issues puts it). Much has been written about the importance of biomimicry in designing the sustainable products and services of the future but here we look at some of the new leadership capabilities emerging in the area of innovation. These can all be seen as examples of 'process biomimicry'.

Developing new sensing mechanisms

All living beings have multiple mechanisms to allow them to sense their environment, so that they can adapt their behaviour to changing conditions. Similarly, good business leaders have always been well-connected to their 'ecosystem' of suppliers, customers, investors, and so on. The ability to sense the mood of the team, the army, the employees is similarly well documented – Tom Peters called it MBWA, Management By Walking About (Peters 2010) but Shakespeare had already documented this approach in Henry V! However, in our more tightly interconnected world new sensing mechanisms are needed, beyond the traditional domain of the business leader and this requires new competencies. Two particular examples are provided by the way WalMart now partners with NGOs it previously considered to be against them and the way Mexican cement producer Cemex engaged with its client base at the 'bottom of the pyramid'.

WalMart has often been attacked for many of its practices, sometimes with good reason. This is not surprising as it is by some measures the biggest company in the world with over two million employees, two hundred million customers visiting its stores each week and a turnover in excess of $400 billion. These criticisms frequently came from environmental NGOs trying to put pressure on WalMart to reduce the environmental impact of its business, whether that be through a switch to sustainable fishing or a reduction in GHG emissions. In response WalMart developed a tough skin, they learnt how to rebut accusations and bad press. But they had no positive story to tell about these issues, nothing to inspire their employees or customers. In the words of former CEO Lee Scott, 'we had a great defence but no offence!' (Scott 2009). Upon realising this Scott started to engage in those organisations which had been the company's biggest detractors, realising that they represented a view of the world which WalMart did not understand and from which they could learn. The results are tangible – WalMart now works closely with environmental NGOs such as WWF, Environmental Defence and NDRC and this relationship has led directly to WalMart's ambitious commitment to reduce GHG emissions described above as well as to a much broader set of collaborations on how to measure and manage reduction in environmental impact.

In the mid 1990s, Mexican cement giant Cemex discovered that their sales to their poorest customers were remarkably resilient to the financial crisis that rocked the country and cut their sales to more affluent customers in half. They issued a 'declaration of ignorance' and sent a team to live in the shanty towns for months with a brief to understand this environment, not to sell more cement! The result of this new 'sensing mechanism' was a profound shift in Cemex's understanding of this part of their market and the development of a radically new program called Patrimonio Hoy (Equity Today) which led to participants building better houses three times faster and at two-thirds the cost than before the program. Note all this improvement in the lives of poor Mexicans has not been bad for business for Cemex! (Hart p.140–44)

Building evolutionary innovation capabilities

The traditional view of innovation in large organisations has been based on the model of well-staffed research and development departments, full of boffins with PhDs dreaming up new products and ideas but here too new models borrowed from nature are starting to take precedence. Strategists have learned that the economic and technological worlds are deeply non-linear. As a result they have largely abandoned attempts to predict the future and form strategies to fit. Instead companies adopt a 'scenario planning' approach where they model several plausible futures and then explore how given strategies would play out in each future (Van der Heijden 2005). Thus they both hone the strategy and their ability to react to unfolding change. These approaches are now given a theoretical underpinning by much of the work on complexity economics, emerging from the Santa Fe Institute and best summarised by Beinhocker. In this model, rather than rational actors maximising utility, businesses are agents in a complex evolving fitness landscape and their ability to rapidly experiment and co-evolve becomes a key to success. This is starting to play out in changing attitudes to innovation. Large firms now recognise that different approaches are needed – ones which follow the evolutionary pattern of innovate, experiment, amplify or kill. This can be seen in the prevalence of internal venture funds, such as Motorola's Early-Stage Accelerator program (O'Connor 2006) and the Wikinomics approach to outsourcing R+D problems to 'the crowd' adopted by giants such as Proctor and Gamble (Tapscott & Williams, pp.103–108).

Building collaborative advantage

Much has been written about the need to build competitive advantage. However, most organisations must cooperate much more than compete and an inability to effectively and creatively cooperate will ultimately lead to a loss of competitive advantage. Why then is so much less written and talked about collaboration than competition? Perhaps because collaboration is more difficult, more complex, more nuanced? It is certainly harder to measure. Collaboration requires dialogue, the constant exchange of ideas which allows a partnership to remain vital, to flex as the world changes and to co-evolve new possibilities. These partnering skills have always been important. In my experience of running manufacturing businesses, for example, companies nearly always have many more suppliers and customers than direct competitors. The relationships with suppliers, customers and employees may all be formally defined by contracts but their health has a lot more to do with the quality of conversations, degree of honesty, trust and generosity. The contract may provide some guidance and a safety net should things go wrong but trust cannot survive an approach limited to the written definition. I remember well the extent to which Toyota helped my old firm when we had a production problem affecting a key component of all their Prius cars. If they had followed the contract by the letter they could have charged us a lot of money for stopping production and switched supply to someone else. Instead they invested a lot of time and expertise in helping us not only to understand this problem but also to improve how we managed our entire production process. The results of Toyota's commitment to collaboration with suppliers over the long term are clear to see and in stark contrast to the typical American or European relationship with more of an emphasis on making suppliers compete with each other.

Leading change beyond the firm

In business the leadership paradigm has been changing for some time. No longer do we look for the hero leader who brilliantly identifies which strategic levers to pull and brings about miraculous change. Organisations are increasingly seen as living organisms (see, for instance, de Geus 1997) or networks of conversations with leaders at all levels influencing the system through each intervention, each conversation. To play their part in bringing about systemic change, business leaders are increasingly recognising the interconnectedness of the whole earth system and seeking to extend their leadership influence beyond the firm or narrow confines of policy affecting their performance. New business leaders recognise that they have responsibility for the whole world. Here I use the word 'responsibility' not in the narrow, linear sense of 'being to blame if things go wrong' but in the sense of 'recognising that everything is interconnected, that every act has consequences and acting accordingly'. I think of this as 'leading beyond the firm'. It requires courage and imagination; courage to start conversations in unfamiliar areas and to risk failure, imagination to believe that the role of business can change. There are many examples of this in recent times.

Influencing regulators

Business leaders have long sought to influence regulation in order to skew markets in their favour. But the language which business leaders are using is changing to reflect the need for policy to change not just for their narrow self-interest but also because of the real and present dangers. They recognise the scale of the challenge and are innovating to thrive in the new reality and so their efforts to influence policy are also to their own advantage. This is a good thing! All partnerships thrive best when there is a clear alignment of interests. A couple of recent examples illustrate this. At a recent seminar at Chatham House in London, Lord Browne (former chairman of BP, now managing director of Riverstone, who manage one of the largest investment funds dedicated to renewable energy, explained the importance of governments implementing clear, long-term policies with regards to energy in order to give investors the kind of clear long-term price signals needed to justify the massive levels of investment we need in order to make the transition to low-carbon electricity generation. Leadership in a resource-constrained world will require the courage to engage with the broader system beyond the firm to press for changing policies, to change the way consumers are educated and to encourage sector peers to join in lobbying for change.

Educating investors

The flow of power between investors and company directors is not one way, it is reciprocal; both influence each other. But both institutions are conservative and this can have bizarre consequences in a rapidly changing world. At the 2010 launch of the Carbon Disclosure Project's results in London where investors and companies discussed the business response to climate change, I witnessed a conversation in which investors complained that they did not know what questions to ask directors. The company directors replied that they did not know what information the investors wanted to receive about their work on climate change. The result? No information, no questions, no conversation! The conclusion was that companies who are taking action should tell investors about this and explain why they think this is making them more competitive (there are plenty of reasons for this – reducing energy reduces costs, developing low-carbon technologies future-proofs sales, the attraction and retention of talent is made easier, regulation is rapidly advancing, considering future climate patterns mitigates supply chain risk and so on).

Telling universities what they need to teach

The next generation of business leaders are at business school right now. They have grown up in a world where the issues of sustainability have become better and better understood. They do not want to commit their lives to being part of the problem. They want to be part of the solution. This has profound implications for educators who are struggling to adapt quickly enough. But change is coming and companies can and should help by more explicitly asking for it. One good example of such change is the recent launch of the 'One Planet MBA' at Exeter University in England, in partnership with WWF (University of Exeter 2011). Fully subscribed in its first year, this program promises to lead the way to Exeter putting sustainability at the heart of all its business education. This is part of a wider trend as documented by The Aspen Institute's annual 'Beyond Grey Pinstripes' survey into the way business schools are integrating issues concerning social and environmental stewardship into the curriculum (Aspen 2011).

I recently attended a workshop on non-financial reporting at Harvard Business School where the new Dean Nitin Nohria made opening remarks acknowledging his institution's role in building the new capabilities for business to lead in solving global problems. As he put it: 'Whether it be the environment, healthcare, or making sure that people have access to information, I can't think of any major problem that society confronts today that can be effectively solved unless business plays an important part' (Nohria 2010).

Conclusions

The institution of business is so big and so powerful and the challenges facing humankind so all-pervasive that it is inconceivable for business to play anything other than a major role in bringing about the scale of societal transition needed to avoid catastrophe. Many who have been warning of the consequences of addiction to consumption and excessive resource use are uncomfortable with this view. They prefer to demonise business, finding comfort in the definition of some 'other' out there to blame. This illusory position fails to meet reality as it is and will delay rather than accelerate solutions because learning will be reduced. Everything is connected. Business does not stand apart but is woven deeply into the fabric of our lives, into the one whole system that we are part of. The complexity scientist Stuart Kauffman describes this whole as 'the great web of biochemical, biological, geologic, economic and political exchanges that envelopes the world' (Kauffman 1995, p.10). At the end of his recent TV documentary 'The Great American Oil Spill' on the aftermath of the BP Gulf disaster, comic and author Stephen Fry summed up his feelings thus: 'It's not an intractable problem. It's a deeply complex one and we're all mired in it and it's a bit of a cop-out to blame one company for being somehow responsible, for offloading our guilt if you like, on to one corporation' (Fry 2010).

The great institutions of business are our creations. They respond directly and indirectly to many signals. We are responsible for them. This is painful to accept when we see the damage caused by the machinery of commerce on a global scale. But this same machinery is constantly changing, sensing its way into innovative new forms, responding to its environment. Business must change and business is changing. Many new leadership capabilities are emerging with increasing examples of implementation, sometimes at staggering scale and pace.

Business leadership is becoming increasingly holistic as managers learn about the realities of our resource-constrained world, understand how to manage the full life-cycle of their products and service, start to incorporate environmental, social and governance issues into their mainstream thinking and seek to build long-term value.

Managers are also learning to innovate in new ways, taking their cue from nature as they develop new sensing mechanisms, build evolutionary ways to create new products and services, and partner to create collaborative advantage

Finally some of the more courageous leaders are recognising that they have a responsibility and a need to effect change beyond the firm, whether in the fields of policy, institutional investment behaviour or education, all affected by the knowledge that our finite world will not support an economic model founded on an assumption of infinite growth.

In recent months two of the world largest consumer goods companies, Proctor and Gamble and Unilever, have announced ambitious goals which would have been unthinkable just a few years ago. P+G have set themselves the target of eliminating all waste to landfill (P+G 2010) while Unilever have committed to doubling sales whilst halving environmental impact (Unilever 2010), precisely the four-fold productivity increase which we need to see across the board.

These commitments are being made ahead of consumer demand and ahead of government legislation because these businesses know that change is inevitable. They will change consumer behaviour and make it easier for governments to introduce regulation. They will lead the systemic change to a global steady-state economy, the only economy compatible with the stark ecological fact that we have limited resources, that we have only this one beautiful planet.

These changes are real. They are happening now. Business is saving the world.

References

Aspen Institute (2011) 'Beyond Grey Pinstripes', located at: http://www.beyondgreypinstripes.org/index.cfm, accessed February 2011.

Beinhocker, Eric (2006) *The Origin of Wealth: Evolution, Complexity and the Radical Remaking of Economics*, Random House, London.

Blood, David (2009) Keynote speech delivered at the launch of the Carbon Disclosure Products FTSE350 results, London, author's notes.

Friedman, Thomas (2008) *Hot, Flat and Crowded*, FSG, New York.

Fry, Stephen (2010) quote taken from documentary 'Stephen Fry and the Great American Oil Spill [4/4]' viewed on YouTube, approximately 11 mins 49 seconds, located at http://www.youtube.com/watch?v=ERWyDb33O88&feature=related, accessed February 2011.

Generation (2010) located at: http://www.generationim.com/strategy/globalequity.html, accessed February 2011.

de Geus, Arie (1997) *The Living Company: Habits for Survival in a Turbulent Business Environment*, HBS, Boston.

Global Footprint Network (2011), homepage found at: http://www.footprintnetwork.org/en/index.php/GFN/page/2010_living_planet_report/ , accessed February 2011.

Hart, Stuart (2005) *Capitalism at the Crossroads*, Wharton, New Jersey.

Van der Heijden, Kees (2005) *Scenarios: The Art of Strategic Conversation*, John Wiley, Chichester.

IIRC (2010) homepage located at http://www.integratedreporting.org/, accessed February 2011.

Kaletsky, Anatole (2010) *Capitalism 4.0: The Birth of a New Economy*, Bloomsbury, London.

Kauffman, Stuart (1995) *At Home in the Universe*, Penguin, London.

Nohria, Nitin (2010) quoted in foreword to *The Landscape of Integrated Reporting*, located at http://hbswk.hbs.edu/item/6532.html, downloaded February 2011.

O'Connor, Jim (2006) *How Motorola uses an early-stage accelerator*, located at: http://www.strategy2market.com/downloads/Visions_Dec06_Motorola.pdf , accessed February 2011.

P+G (2010) located at: http://www.pg.com/en_US/sustainability/environmental_sustainability/index.shtml, accessed February 2011.

Peters, Tom (2010) *Excellence:MBWA* located at: http://www.tompeters.com/dispatches/011777.php accessed February 2011.

Polman, Paul (2010) quote located at http://www.icis.com/blogs/chemicals-and-the-economy/2010/12/unilever-focuses-on-long-term.html, accessed February 2011

Scott, Lee (2009) author's notes from workshop on sustainability metrics with WalMart, suppliers and NGOs.

Smith, Adam (1776, reprinted 1999) *The Wealth of Nations*, Penguin, London..

Sustainability (2009) King III, located at: http://www.sustainabilitysa.org/PressReleases/KingIIIandsustainabilityreporting.aspx, accessed February 2011.

Tapscott, Don & Williams, Anthony (2006) *Wikinomics: How mass collaboration changes everything*, Atlantic, London.

Unilever (2010) as described at: http://www.pg.com/en_US/sustainability/environmental_sustainability/index.shtml, accessed February 2011.

University of Exeter (2011) The One Planet MBA, located at: http://business-school.exeter.ac.uk/mba/, accessed February 2011.

WalMart (2010) located at: http://walmartstores.com/pressroom/news/9668.aspx, accessed February 2011.

Weizsacker, E., Lovins, A. & Lovins, L. (1998) *Factor Four: Doubling Wealth, Halving Resource Use – The New Report to the Club of Rome*, Earthscan Ltd.

17. SLOC: The Emerging scenario of Small, Open, Local, Connected

EZIO MANZINI

In several languages the term 'crisis' has the double meaning of 'risk' and 'opportunity'. There is no doubt that today we currently are facing a deep worldwide crisis. Many people talk about the massive risks we face. But it is also necessary and timely to talk about opportunities generated by the crisis. The opportunities manifest at the intersection of three main innovation streams: *green innovation* driven by the increased evidence of planetary limits, the *spread of networks* driven by new information and communication technologies based on distributed, open, peer-to-peer organisation and the emerging *social economy* driven by the need to tackle very complex social and environmental problems.

This chapter explores these opportunities, considered collectively as the SLOC Scenario, where SLOC stands for Small, Local, Open, Connected. These four adjectives refer to the socio-technical systems on which this scenario is based: a distributed production and consumption system in which the global is a network of locals – a mesh of connected local systems, the small scale of which makes them comprehensible and controllable by individuals and communities.

This chapter outlines how the SLOC Scenario impacts our design cultures which now urgently need to be up-dated if they are to respond appropriately to the crisis. Implementing this scenario demands new visions and proposals, and so clearly designers have an important role to play in generating the SLOC Scenario. They must also act as catalysts of the diffuse physical and social resources needed to make these visions and proposals real.

Cosmopolitan localism

We are discovering that, contrary to what was thought in the past, the joint phenomena of globalisation and networking have given a new meaning to the local. The expression 'local' now refers to something very different from what was meant in the past – that is, the valley, the agricultural village, the small provincial town, all isolated and relatively closed within their own culture and economy. Indeed, the term 'local' now combines the specific features of places and their communities with new phenomena generated and supported worldwide by globalisation and by cultural, socio-economic interconnectivity. Today, these phenomena are often characterised by extremely negative tendencies, that range from traditionalist stances that support local interests, including different forms of fundamentalism hidden behind the protecting veil of traditions and identity (Bauman 1998; Beck 2000) to turning what remains of traditions and landscapes into a show for tourist purposes – a 'Disneyfication' of the local (Bryman 2004).

Luckily, however, we find local communities that are inventing unprecedented cultural activities, forms of organisation, economic models and initiatives that, as a whole, represent an interesting and encouraging possibility which is referred to as *cosmopolitan localism* (Sachs 1998, Beck 2000). In this new trend there is a creative balance between being rooted in a given place and community and being open to global flows of ideas, information, people, things and money (Appadurai 1990, 2001). But this delicate balance that can either tip into a hermetic closure to the outside world or into a openness to outside influences that destroys the particular features of the local social fabric.

The cosmopolitan localism which we are discussing here generates a new sense of place and culture so that places are no longer isolated entities, but become nodes connected to short and long networks. The short networks generate and regenerate the local social and economic fabric, whilst the long networks connect a particular place and community with the rest of the world. In other words, cosmopolitan localism refers to entities that are, *per se,* small and local but that also maintain a distinctive place within the modern global network society.

In networks, small is not small

Some forty years ago E.F. Schumacher wrote his famous book *Small is Beautiful* (Schumacher 1973). At that time, he made a choice in favour of the small and local on cultural and ethical grounds as a reaction to the prevailing trend towards greater scale and delocalisation that he saw around him. Today, we follow Schumacher for these and other, new and compelling reasons. Forty years ago, the 'small' that Schumacher referred to was genuinely small because it had little chance of influencing the larger scale and the local was really local because it was partly isolated from other local communities. Furthermore at that time all the technological and economic trends were moving in the opposite direction, that is, in the direction of 'bigger, and better'. Today the context is totally different since the small can now be influential as a node in the larger global network. The local can now also be open to the global flow of people, ideas and information. In other words, we can say that today the small is no longer small and local is no longer local, at least in traditional terms.

This change in the nature of the small has enormous implications, for better and for worse. Perhap the most potentially beneficial implication is that the global network makes it possible to operate on a local and small scale in very effective ways, because, as we will see in the next paragraph, the networked and flexible systems that emerge provide the only possibility for operating safely in the complex, fast changing, highly risky contemporary environment.

Similar considerations apply to the notion of the local and to the related notion of place. In the last few decades there have been long and important debates about how the globalised flow of goods is bringing about the end of places and localities (Augé 1995; Castells 1996), and it is indeed important to recognise how the flow of goods creates a crisis for traditional places and promotes the spread of homogenised 'non-places'. But these observations do not capture the entire complexity of the new reality where a growing number of people are actively searching for local traditions and for new forms of locality rooted in the modern context of global interconnectivity. Given the new meanings that the terms 'small' and 'local' are assuming in the network society, it is important to consider some of the economic processes that can give cosmopolitan localism a high degree of resilience within the global context .

Local products for global markets and local communities

Within the frame of this cosmopolitan localism a variety of new local, open and highly contemporary activities are taking place: the rediscovery of neighbourhood, the resurgence of local food and of local crafts that carry the spirit and history of a particular place and community to the end consumer, in some cases with worldwide commercial success. This re-discovery of the local as a valuable attribute for specific products in the global market has been until now the clearest evidence of cosmopolitan localism. The most commonly known and quoted examples are some food and food-related products the success of which is strictly linked to their place of origin and the specificity of place in terms of its physical and cultural resources. But the same has been true for some non-food products, such as the essential oils of the Provence region of France, or Murano glassware, in Italy, to give just two examples known worldwide. All these products carry the spirit of their particular place to consumers, who feel a connection to the natural and cultural features of the place from which the products originate.

It has to be emphasised, however, that, to ensure the success of this particular business model, the place to which these products are related needs to be alive, thriving and of high quality. In other words, if products are to the bearers of the spirit of place, the quality of that place, and of the community that inhabits it, must also be guaranteed. Therefore, strong links needs to be established between place, community and product since the quality of the place and of the community are essential for a product's success. Conversely, the long term success of a product needs to promote the qualitative regeneration of place and community.

The success of these local products for the global market is the most visible expression of one important new idea of locality. But, in the last decade, a new and even more important trend has started to appear and evolve: the rediscovery of the value of local production purely for local communities. This growing recognition is based on two powerful motivations: a search for the quality of proximity that comes from the direct experience of the place where a product comes from and of the people who produce it (Petrini 2007), and the search for self-sufficiency in order to promote community resilience to external threats and problems (Hopkins 2009). These ongoing tendencies are very clear in the realms of food and agriculture, where, a variety of initiatives, such as purchasing coops, farmer markets, community-supported agriculture initiatives and, localised, food networks are very effective in minimising food miles.

Symbiotic clusters

Another less well known but very promising possibility for implementing cosmopolitan localism is the clustering of local activities such as multifunctional farms and eco-districts. On multifunctional farms a variety of crops are combined to mutually reinforce each other, mimicking natural biological systems in the spirit of agroecology (Gliessman 2000). Eco-districts are places where, by adopting models from industrial ecology, the wastes of industrial, agricultural or residential activities become valuable material inputs for other activities. In some cases the energy emissions of high temperature processes have been used in other installations with lower temperature requirements such as greenhouses and residential buildings (Graedel & Allenby 1995; Ayres & Ayres 1996, Pauli 1996, 2010).

Both agroecology and industrial ecology are symbiotic clusters where different production and consumption processes cooperate and strengthen one another to reduce an overall environmental footprint. To realise them a new idea of planning has to be developed where notions of place and proximity are crucial. Symbiotic clusters are technically and economically viable only if their different activities are located near to each other: on the same farm, in the same surroundings or in the same district. These partnerships are held together by cooperative links between different partners, and they can also be connected with other clusters creating larger bioregional scale symbiotic clusters.

We have seen how cosmopolitan localism is, or could be, a very favourable context for the creation of ecologically sound socio-technical systems, which could in turn provide effective technical support for empowering the spread of cosmopolitan localism. To better understand this linkage, we now need to introduce the notion of distributed systems.

Distributed systems

The expression 'distributed system' indicates a web of interconnected, autonomous elements capable of accomplishing complex activities (Johansson, Kish, Mirata 2005; Biggs, Ryan, Wiseman 2010). In the last few decades the adjective *'distributed'* has been increasingly used in several different fields. In the realm of information technology we have *distributed intelligence*; in energy systems there is *distributed power generation*; in production processes we have *distributed manufacturing* and in processes of change we speak of *distributed innovation, distributed creativity* and *distributed knowledge*. One classic case – *distributed computing* — has now evolved into *distributed intelligence,* having become mainstream more than two decades ago. Other cases *(distributed power generation* and *distributed manufacturing)* are increasingly important in the international arena.

Distributed intelligence

It is well known that the internet and the increasing sophistication of computing are generating a new form of intelligence distributed within the nodes of large socio-technical systems. Distributed intelligence implies radical changes in our systems of organisation. Until now solid, vertical organisational modes that have been dominant in industrialised society are melting into fluid and horizontal ones as new distributed forms of knowledge and decision-making become more common (von Hippel 2004; Bawens 2006, 2007). Although these ongoing changes are commonly recognised, the direct and indirect implications are not yet totally understood.

Distributed power generation

This usually refers to an energy system mainly based on interconnected small and medium size power generators and/or renewable energy plants. This option has been made possible thanks to the convergence of several factors: the existence of highly effective small and medium size power generators and the possibility of basing these new energy systems on intelligent information networks known as smart grids. Distributed power generation implies radical changes to the dominant idea of what an electrical system is. It also opens up the possibility of a new relationship between communities and their technological assets as well as a more democratic way of managing the energy system (Pehnt *et al.* 2006). Today, even though it is not yet the mainstream energy strategy, distributed power generation is largely recognised as very promising for dense urban spaces and for rural areas in both Northern and in the Southern countries of the world.

Distributed infrastructure

The integration of distributed intelligence and distributed power generation are the pillars of a viable new infrastructure that could also include distributed water management, waste treatment and mobility systems for both goods and people (Biggs, Ryan, Wiseman 2009). Once in place, this distributed infrastructure system could become the technological backbone for new kinds of production and consumption and for a new kind of distributed manufacturing.

Distributed manufacturing

This expression refers to the localisation of production processes so that they are as near as possible to where the product will be used (that is, to the 'point of use'). The main driver here is the search for hyper-lean production, for light and flexible systems capable of delivering customised final products 'just-in-time' and 'on-the-spot' – that is products created specifically for whoever needs them, when and where they are needed. This way of producing, which can be seen as a high-tech version of craftsmanship, already has some interesting applications, from T-shirts, to CDs and books printed on demand in the same shop, to more complex products such as glasses and furniture. The appearance of new technologies such as 3D printers capable of producing complex parts just-in-time and on-the-spot and new modular plants such as the *Fablabs* promoted worldwide by MIT suggest that distributed manufacturing is likely to spread widely. At the same time increasing environmental and transportation costs create powerful incentives that favour these new technologies. The result is that, even if the re-localisation of production is not yet fully with us, it is a viable possibility for several types of product.

Resilience and adaptability

Until now we have seen that the adoption of distributed systems is mainly driven by information and communication technologies that increase connectivity within society, thereby making possible forms of organisation that have previously been very difficult to bring about. The spread of distributed systems is also more likely in a society that is becoming less predictable and ever more turbulent world (Beck 1992; Walker & Salt 2006).

Resilience and risk management

By their very nature distributed systems can more easily deal with failures without collapsing. Thus, they are highly resilient, as opposed to mainstream, highly fragile hierarchical systems (Fiksel 2003; Hopkins 2009). Distributed systems are thus highly appropriate for our high risk society where the risks are environmental, social and political. It is interesting to note that the military demand for highly resilient information systems was the impetus for the research that finally led to the internet as we know it today.

Adaptability and change management

Distributed systems are highly adaptable to changes of context, a feature intrinsic to their networked nature. In fact, different nodes of the network can be easily modified or substituted without interrupting the functioning of the overall system. This flexibility is useful in many circumstances: the economy can adapt its products as the market changes; manufacturing can more easily adapt to client choices and power generation can cope with the variability of renewable resources. The same system adaptability is a fundamental property of distributed systems in their construction phase. In fact, systems based on large interventions need large initial investments and a long time to be implemented, with the risk that, when finally completed, the context might have changed and the technology could have become obsolete. In contrast distributed systems can be more easily implemented using a step by step process with modest initial investments and with a high possibility of modifying them during their implementation as the context changes (Lovins & Lovins 2001; Olsson & Folke 2004).

The SLOC Scenario

Distributed systems and cosmopolitan localism are socio-technical trends driven by different forces and by different people with different motivations. Nevertheless they share two common characteristics: they both seek to foster small connected entities and they are both local and open. Finally, as we have already observed, they are converging on each other and there are good reasons to believe that they will converge and reinforce each other even more in the near future.

The four adjectives in the SLOC Scenario (small, local, open and connected) work well because they generate a holistic vision of how society could be, they are also comprehensible since their meanings and implications can be easily understood by everybody and they are also viable, because they are supported by major drivers of change – that is by the complex relationships between globalisation and localisation, the power of the internet and the adoption of new forms of organisation that SLOC makes possible. If it is well constructed and communicated as a proposal that everybody can understand and discuss, the SLOC Scenario could become a powerful 'social attractor' that, in principle, could trigger, catalyse and orient a variety of social actors, innovation processes and design activities.

Implications and potentialities

The perspective that the SLOC Scenario represents could be interesting purely as a potentially productive line of socio-technical innovation to be explored for its own sake irrespective of its implications for sustainability, but it becomes even more important when one considers its potential contribution to the creation of a sustainable society in terms of environment, society, politics and everyday life.

Environmental implications

The SLOC Scenario is based on the notion of distributed systems. It suggests that agriculture should become multi-functional based on agroecology principles. It would allow energy, water and waste treatment infrastructure to work effectively thanks to small, interconnected plants that use physical resources locally in a sustainable way. In manufacturing,

SLOC production processes can be locally clustered to generate symbiotic industrial ecologies in which the waste of one process is transformed into valuable inputs for other physically adjacent processes. Furthermore, production-consumption chains can be shortened reducing their average transportation intensity and the associated congestion and pollution. In short: the SLOC Scenario is intrinsically coherent with environmental sustainability and can be supported by the same economic forces that are currently promoting green technologies and regenerative economies (Vezzoli & Manzini 2008; Ryan 2009).

Social implications

In the SLOC Scenario a large part of the value creation process takes place at the local scale, as do decisions, and job creation. At the same time, by intensifying local activities and interactions, the SLOC perspective reinforces the social fabric and helps to regenerate a locality's existing social resources. In short: the SLOC Scenario is totally coherent with the criteria of social sustainability and can be enhanced by the same actors that support new ideas of wellbeing, the regeneration of the commons and collaborative social services development (DEMOS 2007; Jegou & Manzini 2008; Kling & Schulz 2009).

Political implications

In the SLOC Scenario decisions are made near the final users. The transparency of decision making processes is increased, thereby facilitating democratic discussions and choices. In particular, individuals and communities are more likely to make responsible decisions given that in the SLOC Scenario the advantages and problems related to decision options can be better compared. In short, the SLOC Scenario proposes a new concept of globalisation where, for each process of production, distribution and consumption, much of the decision making, know-how, and economic value remain in each particular locality in the hands, minds and pockets of the local community (Kanduchar & Haime 2008)

Implications for every day life

The SLOC Scenario proposes small and local everyday life solutions that people can understand and control. Some notable success stories tell us that wherever possible people choose small, local options. A classical example is the success of small personal laptops which thirty years ago were undreamt of in an era dominated by big, impersonal mainframe computers. And, in more recent times, we find the same trend in the growing demand for local food in favour of agro-industrial products. If this observation is correct, we can assume that the SLOC Scenario would be highly preferred if it delivers benefits in socially and culturally acceptable ways that are both small and local and open and connected.

Technical and social innovation

The SLOC Scenario, as we have seen, is both the result of a variety of innovation processes that can in turn catalyse new innovations. These innovation processes are highly complex. They include classic technological innovations ranging from highly efficient centrally produced solar panels and electric batteries to products manufactured locally. But the SLOC Scenario also requires systemic innovation in the very architecture of a given system. It can also require radical innovations that combine existing technologies in unprecedented ways capable of generating brand new functions, processes and meanings.

Systemic changes are, by definition, complex transformations (Geels 2000, 2002; Halen, Vezzoli, Wimmer 2005) that always imply changes in both technical and social components. In some cases technology clearly prevails and the change is mainly driven by the introduction of new technologies such as high performance materials, distributed manufacturing systems and hydrogen fuel cells.

But, as we have already seen, technological innovation need not be the most powerful driver of change – sometimes social innovations prevail. In these cases solutions in response to social demands are conceived and implemented by design experts, but the knowledge and capacities of non-experts are also important. However, social innovation is a very large issue which goes beyond the limit of this chapter (Mulgan 2006, Murray 2009). Here we would only indicate that the SLOC Scenario (and the distributed systems on which it is based) cannot be implemented without considering the dimension of social innovation. Whereas centralised systems can be developed without considering the social fabric in which they will be implemented, this is totally impossible when the technological solution in question is a distributed one. In fact, the more the system is of a distributed nature, the larger is the interface between the new technology and society.

The meaning of this statement can be better understood if we consider the field of distributed intelligence where it is quite evident that the most promising and innovative solutions are based on new technological possibilities conceived and developed by end-users thanks to communication, cooperation and organisation processes ranging from digital platforms for collaborative services in the most mature industrial societies to creative and collaborative use of mobile phones in African villages (Rheingold 2002; Baek 2010). Clearly, then, there can be no SLOC Scenario and no distributed systems without social innovation. Conversely, it will be very difficult for social innovation to grow and to become a powerful agent for sustainable change without a shared vision such as is made possible by the SLOC Scenario.

SLOC Scenario and design

The SLOC Scenario we have outlined here has been introduced as a vision of a possible future resulting from the convergence of several ongoing social and technological trends. But the SLOC Scenario is only a possibility – it points to a desirable future among many other less desirable or even catastrophic outcomes. In other words, the SLOC Scenario is not a forecast of what will happen, but it could become an effective tool for helping to bring about a desirable future.

This short discussion of the SLOC Scenario indicates also that it mainly depends on wide and well articulated design programs. At the same time, given the radical systemic changes it proposes, the SLOC Scenario challenges both what design can do and the meaning of terms such as 'design' and 'designers'. To face this challenge, the first step is to move from the product-oriented design culture of the last century towards one more relevant to this century. This involves a shift from thinking about material products as central towards a systemic approach where the focus is on the interactions between people, things and places. As part of this shift designers have to recognise that the 'objects' to be designed in contemporary society are a complex mix of material and immaterial systems where products, services, people and places are highly interconnected in often unprecedented ways (Meroni 2007; Jegou & Manzini 2008).

A second step is to change the way designers position themselves within production and consumption systems and in design processes. We have already observed that in contemporary society, systemic changes are driven by a growing number of actors who, consciously or not, adopt a design approach (Cottam & Leadbeater 2004a; Leadbeater 2008). In so doing they generate wide and flexible design networks that collaboratively conceive, develop and manage sustainable solutions in which the position and role of professional designers (the design experts) changes. Even if some aspects of this change are still in discussion (see, for instance, Verganti 2009; Brown & Wyatt 2010), it is universally recognised that design experts must collaborate with other partners and that they must consider themselves as active members of these emerging networks, contributing their own competences to improve the overall design capability of the network.

A wide design research program

The adjectives 'small, local, open and connected' are guidelines for triggering and orienting a wide and well articulated design research program in which thousands of design researchers provide their autonomous but convergent contributions to a wide variety of design problems in ways that are truly open and collaborative. Thus the SLOC Scenario architecture and *modus operandi* are similar to the open, collaborative, self-regulating world it hopes to promote.

Who could drive this program? Of course, the whole design community need to take part. A major contribution could come from design schools in universities and colleges. Thanks to the enthusiasm of their students and the experience of their teachers, design schools could become active laboratories where complex problems are tackled, where new visions are generated and where new tools are defined and tested. The potential of design schools for creating a network that could initiate and support social innovation projects and where and experiences could be shared at the local, regional and global scales has so far been greatly under-valued. An international initiative called *Design for Social Innovation towards Sustainability,* has recently been launched to further this aim (DESIS 2010). DESIS is a network of Design Labs, based in design schools, operating as an open access design agency aiming to generate conversations, scenarios and proposals for a truly sustainable future.

References

Appadurai, A. (1990) *Disjuncture and Difference in the Global Cultural Economy*. At: http://www.intcul.tohoku.ac.jp

Appadurai, A. (ed.) (2001) *Globalization*, Duke University Press.

Augé, M. (1995) *Non-Places: Introduction to an Anthropology of Supermodernity*, Varso, London, UK.

Ayres, R.U. & Ayres, L. (1996) *Industrial Ecology: Towards Closing the Materials Cycle*, Edward Elgar Publishers, London.

Baek, J.S. (2010) *A socio-technical framework for collaborative services*, PhD thesis, to be published, Milano.

Bauman, Z. (1998) *Globalization*, Sage Publications, London, UK.

Beck, U. (1992) *Risk Society*, Polity Press, Cambridge, UK.

—, (2000) *What is Globalisation?* Polity Press, Cambridge, UK.

Bauwens, M. (2006) 'The Political Economy of Peer Production,' *Post-autistic Economics Review*.

Bauwens, M. (2007) *Peer to Peer and Human Evolution*, Foundation for P2P Alternatives, at: p2pfoundation.net

Biggs, C., Ryan C., Wiseman J., & Larsen K. (2009) *Distributed Water Systems: A networked and localised approach for sustainable water services*. Victorian Eco Innovation Lab, Melbourne.

Biggs, C., Ryan, C., & Wiseman, J. (2010) 'Distributed Systems. A design model for sustainable and resilient infrastructure,' *VEIL Distributed Systems Briefing Paper* N3, University of Melbourne, Australia.

Bryman, Alan E. (2004) *The Disneyization of Society*, Sage Publications.

Brown, T., Wyatt, J. (2010) 'Design Thinking for Social Innovation,' *Stanford Social Innovation Review*, Winter, 2010.

Benkler, Y. (2006) *The wealth of networks: How social production transforms markets and freedom*, Yale University Press, New Haven and London.

Bernard, J. (1973) *The Sociology of Community*, Scott Foresman, Glenview, IL.

Castells, M. (1996) *The Rise of the Network Society. The Information Age: Economy, Society and Culture*, Vol.1, Blackwell, Oxford.

Cottam, H., & Leadbeater, C. (2004a), *Health. Co-creating Services*. Design Council – RED unit, London, UK.

DEMOS (2007) The Collaborative State. DEMOS, London. At: http://www.demos.co.uk/publications/collaborativestatecollection.

Fiksel, J. (2003) Designing Resilient, Sustainable Systems. *Environmental Science and Technology.* Vol.37, pp.5330–39.

Geels, F. (2002) *Understanding the dynamics of technological transitions: a co-evolutionary and socio-technical analysis,* The Netherlands: PhD University Twente, Enschede.

—, en Kemp, R. (2000) *Transitions from a socio technological perspective* (Transities vanuit een sociotechnisch perspectief). Report for the Ministry of VROM. University Twente, Enschede, and MERIT, Maastricht.

Gliessman, S.R. (2000) *Ecological Processes in Sustainable Agriculture,* Lewis Publishers, London, UK.

Graedel, T.E. & Allenby, B.R. (1995) *Industrial Ecology,* Englewood Cliffs, Prentice Hall.

Halen, C., Vezzoli, C., & Wimmer, R. (2005) *Methodology for Product Service System Innovation.* The Netherlands: Koninklijke Van Gorcum.

Hopkins, R. (2009) *The Transition Handbook: from oil dependency to local resilience,* Green Books, UK.

Jegou, F. & Manzini, E., (2008) *Collaborative services, Social Innovation and design for sustainability,* Polidesign, Milano.

Johansson, A., Kish, P., Mirata, M. (2005) *Distributed economies. A new engine for innovation,* in the Journal of Cleaner Production 2005, Elsevier.

Kanduchar, P. & Haime, M. (2008) *Sustainability Challenges and Solutions at the Base of the Pyramid,* Greenleaf Publishing.

Kling, A. & Schulz, N. (2009) *From Poverty to Prosperity,* Encounter Books, New York.

Leadbeater, C. (2008) *We-Think,* Profile Books, London.

Lovins, A., & Lovins, L. (2001) *Brittle Power,* Book Press, Vermont.

Manzini, E. (2009) *New Design Knowledge,* Design Studies,301.

—, & Jegou, F. (2003) *Sustainable Everyday,* Edizioni Ambiente, Milano.

Mulgan, J. (2006) *Social innovation. What it is, why it matters, how it can be accelerated,* Basingstoke Press, London.

Murray, R. (2009) *Danger and opportunity. Crisis and the social economy,* NESTA Provocation 09, London.

Olsson, P., & Folke, C. (2004) Adaptive Co-management for building resilience in socialecological systems. *Environmental Management,* Vol.34, pp.75–90.

Pauli, G. (1996) *Breakthroughs: what business can do for society,* Epsilon Press, UK. *Svolte epocali* (1997), Baldini& Castoldi, Milano.

—, (2010) *The Blue Economy,* Report to the Club of Rome.

Pehnt, Martin *et al.* (2006) *Micro Cogeneration. Towards Decentralized Energy Systems,* Springer, Berlin.

Petrini, C. (2007) *Slow Food Nation. Why our food should be good, clean and fair,* Rizzoli, Milano.

—, (2010) *Terra Madre. Forging a new network of sustainable food communities,* Chelsea Green Publishing, London, UK.

Ryan, C. (2009) *Climate change and ecodesign*: Part II. *Journal of Industrial Ecology,* Vol.13, pp.350–53.

Rheingold, H. (2002) *Smart Mobs: The Next Social Revolution.* Basic Books, New York.

Sachs, W. (ed.)(1998) *Dizionario dello sviluppo,* Gruppo Abele, Torino.

Schumacher, E.F. (1973) *Small is Beautiful, Economics as if People Mattered,* Blond and Briggs, London, UK.

Tapscott, D. & Williams, A.D. (2007) *Wikinomics. How Mass Collaborations Changes Everything,* Portfolio, New York.

Vezzoli, C. & Manzini, E. (2008) *Design for environmental sustainability.* Patronised United Nation Decade Education for Sustainable Development, Springer, London.

Verganti, R. (2009) *Design-Driven Innovation,* Harvard Business Press.

Walker, B., & Salt, D. (2006) *Resilience Thinking – Sustaining ecosystems and people in a changing world.* Island Press, Washington.

Websites

DESIS 2010, http://www.desis-network.org

SEP 2010, http://www.sustainable-everyday.net

SIX 2010, http://www.socialinnovationexchange.org

18. Wicked Problems and the Relationship Triad

TERRY IRWIN

> All men are designers. All that we do almost all the time is design, for design is basic to all human activity. The planning and patterning of any act toward a desired foreseeable end constitutes the design process. Any attempt to separate design, to make it 'a thing' by itself works counter to the fact that design is the primary underlying matrix of life.
>
> VICTOR PAPANEK, *THE GREEN IMPERATIVE* (1995)

Solutions arise out of the design process – we are all designing, all the time. The transition to a sustainable society is one of the biggest design challenges the human race has ever faced. Meeting this challenge will require countless designed solutions that will be created by people from all walks of life, using design thinking and design process. *Design affects and concerns everyone.*

Design is directly linked to the satisfaction of human needs. You might even say that design is an 'emergent' property of humans striving to meet their needs. These needs can be as simple and mundane as planning an evening meal, or as large and consequential as the redesign of an urban transportation system or a course of treatment for a desperately ill patient. Society's transition to a sustainable state is a challenge for a new breed of 'transition designers' working within a new design paradigm, across disciplinary and professional divides.

Myriad problems confronting society in the twenty-first century need to be addressed – global warming, pollution, water shortages, the global economic crisis, poverty, terrorism, the widening gap between rich and poor, to name a few. These, and others like them, can be considered 'wicked problems', a term coined by twentieth century planner Horst Rittel (1972, 1973, Buchanan & Margolin 1995, pp.3–20, Protzen &

Harris 2010, pp.155–65) to describe a type of ill-defined, complex, systemic and purportedly unsolvable problem. Such problems are comprised of seemingly unrelated, yet interdependent elements, each of which manifest as problems in their own right, at multiple levels of scale. *The ability to solve wicked problems will call for new ways of thinking about design, our world and the human presence in it.*

Einstein famously said that problems cannot be solved within the same mindset that created them, therefore solutions to the complex, interdependent problems confronting society in the twenty-first century must arise out of a new mindset or worldview upon which a new design paradigm can be based. The fundamental principles of design can be applied from within every discipline, by people from all walks of life, making transdisciplinary collaborative design a crucial skill (or discipline in its own right) in the coming decades.

The central premises of this chapter are:

1. that wicked problems and their contexts are complex systems that operate according to the same intrinsic principles and dynamics as living systems;
2. that these systems are comprised of countless strands of relationships between people, the environment and the things that people make and do – a relationship triad;
3. that these principles have the potential to inform a new kind of design process that will be better equipped to address wicked problems; and
4. that a new mindset is needed, one that enables people to see wicked problems and conceive fundamentally different solutions which incorporate ethics and a deep concern for both the social and environmental spheres.

Wicked problems

Although the design disciplines have been aware of 'wicked problems' since Rittel identified them in 1973, and have developed myriad processes and strategies for addressing them, efforts have not been directed toward understanding their dynamics and anatomy which is crucial to developing meaningful solutions to them. Discoveries in mid-twentieth century science, which include principles such as chaos and complexity theories, emergent properties, self-organisation, and so on (Goodwin 1994, Briggs & Peat 1989, 1999, Augros & Stanciu 1987, Capra 1996, Wheatley 2006) can provide keys for both understanding wicked problems and transforming design process (see Figure 1). Rittel's theory distinguishes between 'tame' and 'wicked' problems and he was one of the first design theorists to maintain that traditional linear, cause and effect design processes were inadequate for solving complex wicked problems. Due to their multiple levels of complexity and stakeholders with opposing viewpoints or worldviews, Rittel argued that it is impossible to arrive at a complete or fully correct design solution – the problem continually mutates and evolves.

1. There is no definite formulation of a wicked problem.	The information needed to understand the problem depends upon one's idea for solving it. In order to describe a wicked problem in sufficient detail, one has to develop an exhaustive inventory for all the conceivable solutions ahead of time.
2. Wicked problems have no stopping rules.	In solving a 'tame' problem, designers know when the solution has been found. With a wicked problem, one can never arrive at a complete or fully correct solution. The problem continually mutates and evolves.
3. Solutions to wicked problems are not true-or-false, but better or worse.	The criteria for judging the validity of a solution for a wicked problem are strongly stakeholder dependent. However, the judgments of the different stakeholders are likely to differ widely. Different stakeholders see different solutions as simply better or worse.
4. There is no immediate and no ultimate test of a solution to a wicked problem.	Any solution, after being implemented, will generate waves of consequences, over an extended – virtually unbounded –period of time. Moreover, the next day's consequences of the solution may yield utterly undesirable repercussions which outweigh the intended advantages or advantages accomplished so far.
5. Every solution to a wicked problem is a 'one-shot' operation; because there is no opportunity to learn by trial-and-error, every attempt counts significantly.	Every implemented solution is consequential and ramifies through the system. It leaves 'traces' that cannot be undone. And every attempt to reverse a decision or correct for the undesired consequences poses yet another set of wicked problems.

Figure 1.Characteristics of wicked problems. Based on Rittel 1972, 1973, Protzen & Harris 2010, pp.155–65.

A barrier to 'transition design' is the inability of designers and their collaborators to see and understand these types of complex, systemic problems. Global terrorism is an example of a wicked problem whose solution has been attempted by countless governments and their agencies worldwide. Because of the number of stakeholders involved with opposing beliefs and value systems, the intricacies of the political relationships involved, the implied struggle for natural resources such as oil, and the large disparities of wealth between countries, groups and individuals, to name just a few, any attempt to solve part of the problem, affects the dense web of relationships of which it is comprised (Capra 2001). Any solution imposed upon this web or complex system, will have consequences that ramify throughout it in unpredictable ways, at multiple levels of scale. *There is no single correct solution that can be designed.* Simply learning to *see* the complexities and interdependent relationships that comprise a wicked problem such as this is a wicked problem in its own right.

Traditional design processes are based upon linear/reductionist cause and effect models (see Figure 2) aimed at solving tame problems that have well-defined problem statements which lead to clearly 'right' or 'wrong' solutions (Archer 1984, pp.57–82, Simon 1988, p.148). This one-size-fits-all approach is applied to a wide range of problems and is characterised by the limited context within which each problem is viewed. Simplifying the problem contextually achieves two important goals: first, it simplifies the problem enough to make it 'solvable'; and second, simplification makes it possible to solve problems quickly and cost-effectively (see Figure 3a).

Business propositions that are conceived within the dominant, single-bottom-line economic paradigm (Korten 2000, Gray & Milne 2004) with profit as the primary objective, will almost always generate design briefs for solving tame problems. The stated objective of the design process is to arrive at the most cost-effective and profit-generating solution via the most expeditious route. The majority of professional designers and business people currently work within this model. When financial concerns take precedence over social and environmental ones however, the tame solutions generated often exacerbate the wicked problems which from the greater context for the problem. Because of the narrow context established in the design brief and the solitary focus on profit, this connection goes unseen, the larger problem worsens, and the cycle continues.

Typical Phases	Typical Steps	Objectives
1. Find out about/ understand the perceived problem or 'need'	Research/ exploration analysis of findings Formulation of design brief	Designers and researchers attempt to understand both the problem and its context. They analyse their findings in order to frame a clear problem within an appropriate context.
2. Formulate ideas for its solution	Develop concepts Design the preferred solution Prototype/ test the solution	Based upon the 'design brief' formulated in phase one, designers explore multiple ideas/ concepts. The most viable of these is developed further and often a rough prototype of some kind is developed to 'test' in controlled situations.
3. Do or make something	Implement the final solution	Once a final solution has been designed and agreed, designers usually hand it off, or collaborate with others to implement the solution (manufacturers, developers, programmers, fabrication experts, etc.).
4. Evaluate and revise	Reiterate based on feedback	This phase is costly and in traditional design processes (software is one obvious exception) is often abbreviated or eliminated altogether.

Figure 2. Archetypal steps in the design process. For examples see Archer 1984, Jones 1980, Protzen & Harris 2010.

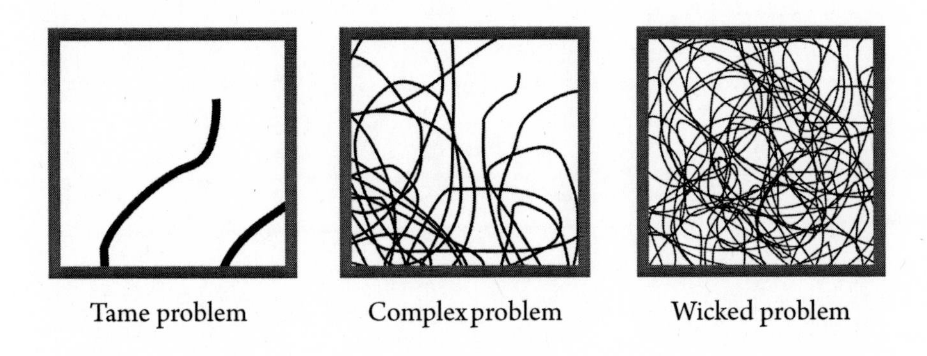

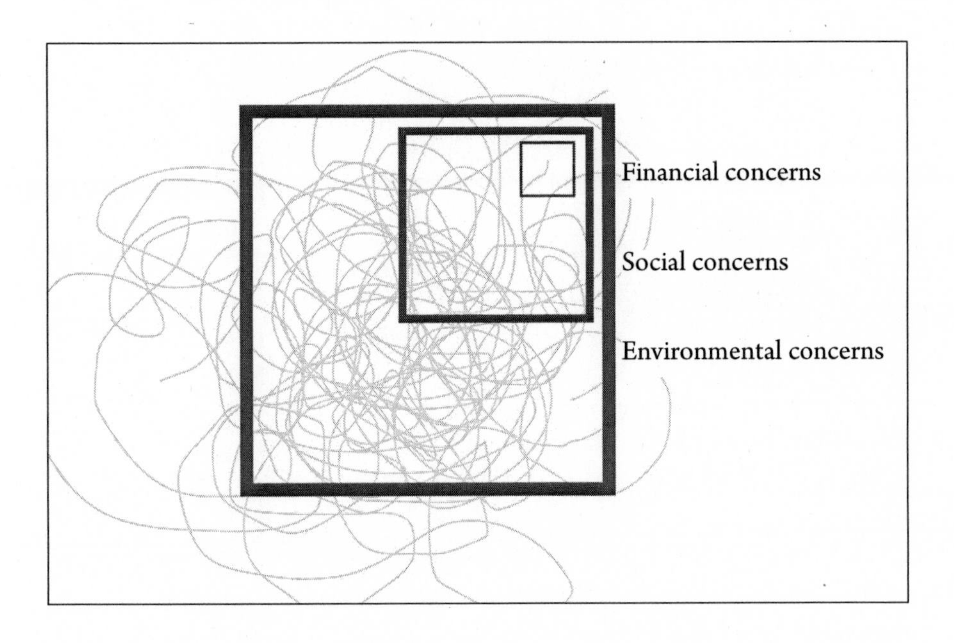

Figures 3a & 3b: These diagrams represent three problems of varying complexity, framed within a context. Figure 3b shows that they are all part of one larger 'wicked' problem and that the perceived degree of simplicity or complexity depends upon how tightly the problem is framed within a particular context.

Design solutions manifest in several broad categories: messages, products and artifacts, processes and built environments are all examples of the work of designers or groups of professionals using the design process to develop solutions. How a design problem is originally perceived and framed is the critical point that determines whether a solution will be sustainable or unsustainable. And it is at this critical point that it will be perceived as either a wicked or a tame problem.

To illustrate an example of a seemingly tame design problem, imagine a business proposition to market a new brand of bottled water. The client would typically engage a designer to develop packaging and would present him with a set of criteria or objectives such as:

—the package must be designed and manufactured with a certain price-point in mind;
—it must not be over a certain weight or shipping/distribution costs will be affected;
—marketing studies may suggest certain user/ergonomic preferences that need to be incorporated into the design; and
—the deadline is very tight, which reduces the time available for exploring alternatives for materials and manufacturing processes.

These criteria are aimed at making the bottle of water as desirable, usable and cost-effective as possible in order to maximise profits for the company and its shareholders. Within such a simplified and controlled context, the designer is solving a tame problem. If the bottle is fabricated from cost-effective materials, holds/delivers water successfully and is given a desirable shape that makes consumers want to buy it, the design is deemed successful because the single objective of generating profit has been met. The designer has solved a tame, straightforward problem.

If, however, social and environmental concerns are incorporated into the design brief (either by the client or at the suggestion of the designer), the level of the problem's complexity increases exponentially and its interdependencies to larger 'wicked' problems will be revealed. When taking into account social concerns, the designer would need to ask different questions: will the plastic chosen for the package leach harmful chemicals into the water and/or will it bio-degrade toxically in a landfill when its useful life is finished? Are the workers who help manufacture or distribute the bottle working under adverse, harmful or exploitative conditions? With respect to environmental concerns, will the plastic from which the bottle is fabricated involve the exploitation, destruction or waste of natural resources (many plastic bottles are petroleum-based)? Does the distribution system involve a high embodied energy cost (how many miles does the bottle travel from source to manufacture to destination)? Will the ultimate demise of the plastic bottle harm the environment and the eco-system within which it will bio-degrade? If the designer follows these lines of inquiry, he will quickly see that they lead to larger wicked problems such as (ironically) the pollution of regional ground water from toxic landfills, health issues posed by toxins in all manner of plastics (Imhoff 2005), the global water shortage and ultimately the global oil crisis, which itself is related to global terrorism. These relationships to larger, wicked problems which exist at ever larger magnitudes of scale are ever present, however the problem must be reframed before they are revealed (see Figure 3b).

The important point is this: *any problem becomes wicked when social and environmental concerns are taken into account.* Tame problems are almost always illusory; they are poorly framed fragments of wicked problems and designers fail to see wicked problems and, moreover, do not understand the dynamics at work within them.

Living systems principles and the relationship triad

Complex systems such as regional watersheds, corporations, our bodies, the WTO, and cities and wicked problems such as terrorism, childhood obesity, and global warming all comprise dense webs of relationship and operate according to the same principles that can be observed within ecosystems (see Figure 4). The relations within these systems fall into three categories; the relations between people, between people and the things they make and do (design), between people and the environment and between the things people make and do and the environment (see Figure 5). If designers learn to understand and leverage these dynamics to develop webs of relationship that are mutually beneficial instead of detrimental (as they are in wicked problems), they will by default, begin to design more responsibly and sustainably.

Characteristic	Description	Revelance for design
Organisationally open, self-organising and 'autopoetic' *(examples of open systems include an individual, our respiratory system, a social organisation such as a community group, university or company, an ecosystem or the planet)*	Living systems are structurally closed and organisationally open and are in a continual exchange of energy and matter with their environment. They are 'self-making' or autopoetic. These systems are self-determining and respond to perturbations from the environment through changes in their behaviour and form. In this way, open systems couple with their environment and co-evolve in symbiotic relationship. Change within such a system can be perturbed, but its outcome cannot be predicted. The more freedom a system has to self-organise, the more creative forms of order will arise within it, and the more resilient it will be to radical changes in its immediate environment.	Designers almost always design *for* and within open systems. Understanding their intrinsic dynamics is key to formulating meaningful solutions. Because these systems are self-determining, any design solution can be seen as a 'perturbation' from its environment. Therefore, the outcome of any design solution cannot be predicted. This suggests a slow, thoughtful process of iterative, user-centred prototypes in which continual 'feedback' informs subsequent iterations. Designers must look for signs of solutions already present in the system (grassroots) and work to amplify them. Solutions designed out of context and 'imposed' upon the system almost never work for long.
Operate in a state far from equilibrium *(in linear, closed systems, equilibrium is the point at which the capacity for change has been exhausted; it has run out of energy)*	Living systems maintain themselves in a state far from equilibrium (a system in equilibrium or stasis would cease to be alive) due to their constant exchange of energy and matter with their environment. By importing energy from outside, these systems keep themselves off-balance, which results in growth and change.	The contexts for most design solutions are open systems, which can appear chaotic and 'disordered'. Traditional design process compels designers to 'restore' or design order (equilibrium) within the system. However, 'disorder' is often a rich bed out of which new forms of order can arise.

Characteristic	Description	Revelance for design
(cont.)	*Disequilibrium is a necessary condition for evolution.* In the face of external disturbances, living systems posses the ability to reorganise themselves and adapt through new forms of order and behaviour.	Instead of imposing order upon the system in the form of a design solution, designers can shift their role to that of 'catalyst for change'. They can look for signs of new behaviour/ forms of order and help amplify them, or usher them into existence. This represents a radically different posture than traditional design process elicits.
Emergent properties	Emergence refers to the new forms of physical and behavioural order that arise within complex systems at critical points of instability and in response to perturbations in the environment. This order is predominantly unpredictable. Complex systems evolve as nested and hierarchical 'levels' which exhibit laws and properties that do not exist at other levels. These new behaviours are said to be 'emergent'. This means that complex systems cannot be understood through reductionist analytics in which the sum of the individual parts explains the behaviour of the whole and in which the whole can be understood independent of the context within which it exists.	It is important for designers to understand the principle of emergence and looks for signs of it within the systems they are designing for/within. In some ways, design is the antithesis of emergence. If a design solution is too rigid, too large in scope and does not take into consideration 'local conditions' (if it is a 'one size fits all' solution), then sooner or later, new forms of behaviour and physical order will arise in response and perhaps opposition to it. This is a sign that the solution is somehow in dissonance with the dynamics of the system. The principle of emergence suggests that designers carefully study the dynamics of the system first, look for emergent properties and design solutions that leverage these tendencies.

Characteristic	Description	Revelance for design
Networked, holarchic, fractal structure	Living systems are comprised of de-centralised, networked, interdependent parts connected through symbiotic relationships. They are holarchic; meaning they are comprised of systems nested within other systems (atoms, cells, organ systems, organisms, communities of organisms, ecosystems, etc.). Individual parts form wholes that are themselves parts of greater wholes. As systems evolve, new forms of emergent order arise as new levels within the system are formed. These systems are often fractal, meaning they are self-similar at multiple levels of scale.	Framing design problems within appropriate contexts is crucial to developing appropriate solutions. If designers design for and within complex systems, then those solutions will be implemented with multi-level, holarchic structures comprised of webs of relationship. Therefore, it is incumbent upon designers to look for the holarchic, networked relationships that comprise the problem and which will affect any solution designers formulate. An understanding of this structure also forms a caution toward design solutions that are centralised, linear and monolithic. It suggests design for relationship as opposed to design that is focused on physical artifacts.
Sensitivity to initial conditions	Living systems often display extreme sensitivity to initial conditions; small changes have the potential to create large variations in the long-term behaviour of a system. Examples are global warming, weather patterns, a viral epidemic. This principle from chaos and complexity science posits that the deterministic nature of these systems (indicated by their initial conditions) cannot be	Solutions are almost always designed for the initial conditions of a particular situation (context), which could be considered a 'snapshot' in the life of a complex system. Solutions are usually designed without taking into account the systems sensitivity to initial conditions and the likelihood that small changes (even the introduction of a design

Characteristic	Description	Revelance for design
(cont.)	predicted due to the chaotic behaviour of the system over time as it is perturbed by small and large changes in conditions. Long-term predictions of its behaviour are impossible.	solution itself) can ramify exponentially throughout the system, producing unexpected results. This suggests that designers acknowledge that solutions built on a large scale, through powerful technologies, can have unpredictable and dangerous effects. Conversely, small changes in behaviour and the introduction of new ideas and practices within complex social systems can likewise ramify exponentially to create sweeping powerful change.
Feedback	Within living systems, single events are part of circular, cause-and-effect chains which affect present and future functioning of the entire system. Negative feedback loops stabilise the system (an example is a thermostat on a heating/cooling system), while positive feedback loops amplify/exacerbate changes (feedback on a microphone). Living systems maintain their ability to survive and adapt through the presence of multiple feedback loops that restore balance to the system after a large disturbance. These same principles are at work	Once designers understand the principles of feedback loops they can be assessed as a critical aspect of solutions development. If change/growth/evolution are integral to the solution, then harnessing the power of positive feedback through the introduction of new information, behaviours, etc. can be designed. If a greater degree of order or stabilising behaviour is connected to the desired solution, then designing around negative feedback loops can be a powerful design tool. Feedback loops are intrinsic to

Characteristic	Description	Revelance for design
(cont.)	in social systems and are at the heart of both static and chaotic situations. The more disturbances a system responds to, the greater degree of flexibility and resilience it develops and the more dynamic (creative) it becomes.	communication systems and can be a powerful way to design solutions.
Diversity	In living systems, diversity refers to the multiplicity of relationships or variables with it. The greater the degree of diversity with a living system, the more flexible and resilient it is. The health of an ecosystem is almost always related to the degree of diversity found within it.	In their desire to bring order to chaotic or problematic situations, designers often impose uniformity upon situations. Solutions are often mass produced with the assumption that one-size-fits-all. Such solutions fail to incorporate diversity and redundancy and are often not suited to local conditions. This suggests a design process that takes into consideration local conditions, incorporates a diversity of relationships/variables and which is designed for ongoing change.
Non-linear relationships	Living, 'open' systems are comprised of non-linear relationships in which causes do not produce predictable, proportional effects. Non-linear relationships are closely related to the concepts of feedback and sensitivity to initial conditions and explain why systems can	This principle is closely related to feedback and networked/holarchic structure. Once designers understand the nature of non-linear relationships, they can be leveraged to introduce new information into the system (communication design) to create learning

Characteristic	Description	Revelance for design
(cont.)	behave in counter-intuitive ways. Non-linear feedback processes are the basis of the instability/ disequilibrium in systems, which often give rise to new behaviour.	networks. The non-predictability of these non-linear, non-causal relationships also holds a caution for designers – the effects and ramifications of design solutions introduced into a system may have unexpected positive or negative results.
Whole and part	Living systems are comprised of semi-discrete 'wholes' which exist at multiple levels of scale (holarchies). The whole has properties that are irreducible to those of its individual parts. Conversely, characteristics of the whole can always be found within its individual parts and provide a clue to the 'essence' of the whole.	A fundamental step in designing solutions is drawing a boundary around a particular problem to be solved; establishing a context. This principle tells designers that this boundary will always be arbitrary and that the 'whole' that it establishes is in a sense 'counterfeit'. Design solutions will exist within contexts that are at once wholes and parts of greater wholes. It is important when examining the context for solutions, to try and understand this principle. The 'parts' designers examine and assess as context for a solution will provide clues to the greater 'whole' of which they are part. This ability to see context and solution at multiple levels of scale is key to the design of sustainable solutions.

Characteristic	Description	Revelance for design
Interdependence and cooperation	Living systems, particularly ecosystems, are interconnected via a vast network of reciprocal relationships. These relationships enable the system to withstand perturbations from the environment and change in response to them (flexibility and resilience). The system can be seen as a part within a greater whole, and in this way, the health and evolution of part and whole are mutually determined. Similarly, the parts within a system are also connected through webs of symbiotic relationship; a web of life.	This principle underscores the need for designers to focus on the webs of relationship that comprise the systems within which they are designing. Understanding the nature and interdependence of these relationships and being mindful of how small changes can ramify through a system, can influence the design of a solution. Part of any design brief can be the objective of creating, fostering and enhancing symbiotic relationships between components/ members of a system. Designers must also remember that often the web of interdependent relationships within a living system is so complex, that it is beyond the capability of a designer or design team to predict the outcome of a solution. This suggests sensitivity, humility and a highly iterative process in the design of solutions.

Figure 4. (Begins p.240) Living systems principles and their relevance to design. Based on Augros & Stanciu 1987, Briggs & Peat 1989, 1999, Capra 1996, Goodwin 1994, Wheatley 2006.

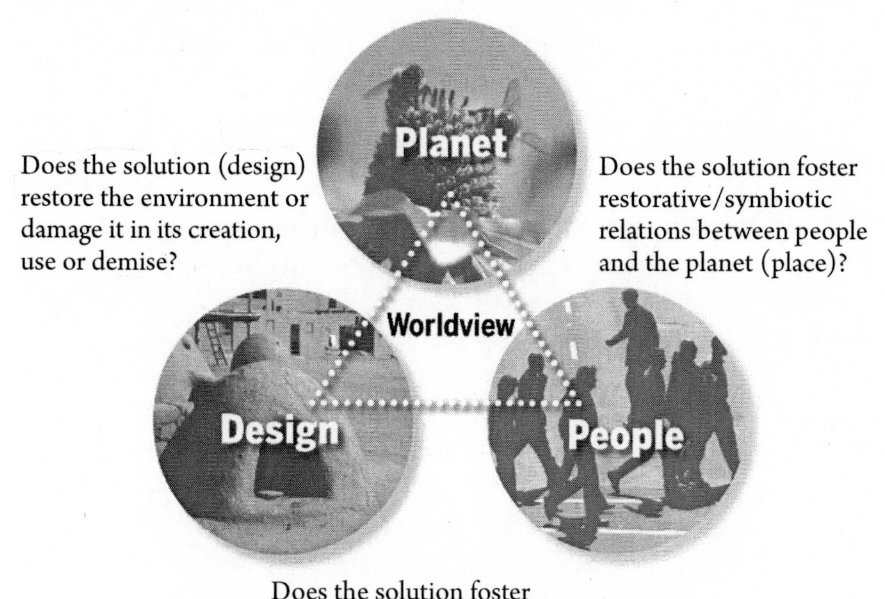

Figure 5. All wicked problems can be seen as complex systems comprised of strands of relationships between people, between people and the planet, between people and the things they do and make (design) and between the things people do and make and the planet. This model can be used as a tool in design process to ensure that social and environmental concerns are integrated into solutions as they are conceived and that these concerns are present from the creation through to the demise of a solution.

It may seem counter-intuitive to suggest that highly complex systems and wicked problems are comprised of just three types of relationships, however the number and nature of these relationships is limitless and highly nuanced which accounts for their complexity and sensitivity to initial conditions. When mapped, these relationships resemble complex food webs found in ecosystems (Yoon, Yoon, Martinez, Williams & Dunne 2005) that are composed of mutualistic interdependencies between species. Unlike an ecosystem, however, the triadic relationships found within wicked problems are most often *not* reciprocal, nor are tight feedback loops present which provide checks and balances within healthy ecosystems. The relationship triad can be used as a tool in the design process to help analyse and reconstitute these relationships and serve as a reminder during the design process that social and environmental concerns must be integrated into solutions along with concerns for profit (triple-bottom-line economics).

Similarly, the dynamics at work within these systems can also become a powerful tool for designing for transition and an understanding of principles such as emergence, feedback and sensitivity to initial conditions can influence the way in which problems are perceived, framed within a context and addressed. These and other principles listed in Figure 4 suggest that the consequences of short-sighted, uninformed or hasty design decisions are likely to ramify throughout these systems at the local, regional and even global levels in unpredictable ways. Chaos and complexity theorists often cite the metaphor of a butterfly flapping its wings in the Amazon forest triggering a hurricane in southeast Asia to demonstrate the power and sensitivity of complex systems.

The Mexican Gulf oil spill of 2010 is an example of a problem in need of a design solution. We don't usually think of design in these terms, but the team of scientists, engineers and executives that assembled needed to collaboratively design a solution to one of the worst environmental disasters in modern history. It was a problem that traditional design processes were inadequate to address. If the oil spill was the problem, the application of dispersants was one of the earliest design solutions.

British Petroleum and the Coast Guard's use of dispersants are increasingly seen by scientists as having been a hasty/premature measure and it is a good example of a wicked problem that was approached as if it were tame. A *Washington Post* newspaper article speculated:

> Now, scientists say, it's difficult to tell what the added use of dispersants permitted by the Coast Guard meant for the gulf. The chemicals may have helped break up some oil before it reached sensitive marshes along the Louisiana coast. But it also may have poisoned ecosystems offshore, helped deplete underwater oxygen and sent oil swirling through the open-water habitats of fish and coral.
> (Fahrenthold & Mufson, August 1, 2010)

The visible oil slick was treated as if it was a tame problem and was framed within a tight context so that an immediate solution could be designed; liberal application of dispersant, whose objective was quick results. Many scientists now argue that this action may have exacerbated the effects of the oil spill in ways that cannot be predicted and may never be fully understood. Some of the consequences of this hastily implemented design solution will undoubtedly include the loss of marine life upon which a large percentage of the Gulf region's economy is based. Still other scientists predict that the oil spill will have both short and long-term negative effects on the health of the regional human population in the Gulf area (Salahi, Allen & Trueger 2010).

The oil spill itself was a wicked problem, but it was/is also part of larger wicked problems at higher levels of scale, such as America's energy policies, its high-consumption lifestyle (*Ecological Footprint Atlas* 2010), global warming (which is connected to petroleum-based fuels) and the pollution of local, regional and global ecosystems, to name a few. Wicked problems are often fractal in their structure; analysis of the problem at a local level reveals connections and interdependencies to more complex problems at higher levels of scale. Solving for a part of the problem at one level of scale will always have consequences that ramify up and down to other levels within the system which can have unintentional adverse effects if system structure and dynamics are not understood prior to introducing a design solution. Traditional design process and our dominant linear/reductionist mode of thinking blinds us to these considerations and effects.

If the teams of scientists and engineers responsible for designing the solution to the surface oil slick in the gulf had considered the problem within the temporal context of the entire gulf ecosystem, it is unlikely they would have identified dispersant as the best solution. These systems principles therefore, hold both a caution and a promise for transition designers; when systems (and wicked problems) are perturbed in ignorance by business propositions and designed artifacts/solutions, sweeping and unpredictable negative consequences can result. If however, these principles are understood they can be 'leveraged' as part of the design solution, by harnessing ramifying effects toward positive change.

Systems theorist Donella Meadows (Meadows 2008) identified what she termed 'leverage points' within complex systems that have the potential to be integrated into the design process. Her seminal article 'Places to Intervene in a System' lays out twelve leverage points for creating and directing change within a system, in increasing order of effectiveness (see Figure 6). Meadows' leverage points are simply a strategy for working with the intrinsic dynamics of complex systems, much as a martial artist uses the momentum of his opponent's action to his advantage. An important distinction within this framework is the idea that some leverage points or places of intervention are more powerful and effective than others. For instance, tinkering with the numbers or the amounts of variables within a system is not as effective as changing the goals of the system. To return to the example of the Gulf oil spill: reducing the *amount* of dispersant that was released into the Gulf would certainly have lessened the negative impact on marine life and the potential health risks. However, questioning the advisability of its use would have been better. If the *goal* of the design solution had included social and environmental concerns (short, mid and long-term health of the ecosystem and regional population), then dispersant would have been ruled out as a viable design solution. Changing the 'goal' or objectives for the design solution is a more powerful leverage point for change in the context of this wicked problem, but would require a decidedly different way of thinking about both the problem and solution.

12. Numbers: Constants and parameters such as subsidies, taxes, standards.	Quantities and amounts and the amounts of flows. Doing more or less of something. Changing these variables is a small fix because it doesn't change the behaviour of the system.
11. Buffers: The sizes of stabilising stocks relative to their flows.	A physical entity used to stabilise a system. Examples are the buffer of money in the bank, an inventory of stock. Because they are physical, they are difficult to change.
10. Stock-and-Flow-Structures: Physical systems and their nodes of intersection.	The physical structure in a system that directs how things move/flow. It is important to get the design right because it is difficult to change once in place. Understanding the limitations and maximising efficiencies is the best way to work with stock and flow structures.
9. Delays: The length of time relative to the rates of system changes.	Delays in feedback are critical determinants of system behaviour. A system can't respond to short-term changes when it has long-term delays. Not a strong leverage point for change because change the speed/rate of delays is difficult to change – things usually take as long as they take.
8. Balancing feedback loops: The strength of feedbacks relative to the impacts they are trying to correct.	A more powerful level that involves information and control as opposed to physical structures. A system usually has several balancing feedback loops. Its strength depends upon the accuracy of and rapidity of monitoring, the quickness

(cont.)	and power of response, the directness and size of corrective flows. Examples are preventative medicine (exercise, good nutrition, and so on). Freedom of Information act to reduce government secrecy, pollution taxes, etc.
7. Reinforcing Feedback Loops: The strength of the gain of driving loops.	In contrast to a balancing feedback loop, a reinforcing feedback loop gains more power the better it works; the more money you have in the bank, the more money you earn. Reinforcing feedback loops are sources of growth, explosion, erosion and collapse in systems. A system with an unchecked reinforcing feedback loop will destroy itself. Reducing the gain around a reinforcing feedback loop (slowing growth) is usually a more powerful leverage point in systems than strengthening balancing feedback loops.
6. Information Flows: The structure of who does and does not have information.	Delivering feedback to a place where it wasn't going before. Missing information flows is one of the most common causes of system malfunction. Incorporates accountability into a system, which is why it is often unpopular.
5. Rules: Incentives, punishments, constraints.	The rules of the system define its scope, its boundaries, its degrees of freedom. Rules represent the power to control behaviour within the system. To understand the deepest malfunctions in systems pay attention to the rules and who has power over them.
4. Self-Organization: The power to add, change, or evolve system structure.	The ability to evolve or change the system at any of the lower levels; changing physical structures, adding feedback loops, changing the rules, and so on. This leverage point implies change, which can be unpopular and uncomfortable.
3. Goals: The purpose or function of the system.	The goal(s) of the system direct all the lower levels of the system. Although the goals direct all the other components of a system, they are not always apparent to everyone operating within the system. The person or group with the power to direct the goals of the system, has real power.
2. Paradigms: The mindset out of which the system arises.	Example: The shared idea in the minds of society, the great big unstated assumptions, constitute that society's paradigm or deepest set of beliefs about how the world works. Paradigms are the source of

(cont.)	systems. From shared social agreements about the nature of reality come systems goals and information flows, feedback, stocks, flows, and so on. Ironically, paradigms can be quick to change: a click in the mind, a new way of seeing.
1. Transcending paradigms.	The ability to keep oneself unattached in the arena of paradigms, to stay flexible, to realise that no paradigm is 'true', that every one, including the one that shapes one's own worldview, is a limited understanding of a world beyond human comprehension.

Figure 6. Places to intervene in a system, in increasing order of effectiveness. Based on Meadows 2008, pp.145–65.

Leverage Points	**9. Quantities**	**8. Buffers/physical structures**	**7. Stocks & flows**	**6. Feedback**
Design Approach	Design for greater efficiency/less waste	Design with bio-degradable/non-toxic materials	Redesign products as flows of services	Design for iteration: incorporating feedback
Existing Design Methodologies and Processes	• Hierarchy of waste management • Factor X eco-efficiency • LEED • Biomimicry (narrow)		• Product Service Solutions (narrow) • Bespoke Product Services • Industrial Ecology (narrow) • Service design (narrow)	• Dynamic lifecycle analysis • Cradle to cradle • Design for disassembly
	Changing Designed Artifacts			
Examples of 'Enabling' Government Policies	WEE, ELVs, Battery take-backs, ROHS Directives Community recycling			IPP and EuP
Opportunities for Design & Designers	Design to improve resource/energy efficiency and reduce waste	Designing with new, non-toxic/bio degradable materials, designing for greater durability/longevity	Working with business to develop service models and ownership solutions that conserve/extend resources	Design for the entire lifecycle of artifacts from conception to demise

Figure 7 shows a diagram in which several of Meadows' leverage points have been mapped on a continuum along with design processes and government policies which support them to demonstrate effectiveness relative to system/wicked problem dynamics. The diagram suggests that it is possible to work all along a continuum of effectiveness, and shows areas in which new design processes need to be developed. Most importantly, it shows that the greatest potential to design sweeping change lies in shifting paradigms (worldviews) and lifestyles; what we believe/care about influences how we perceive problems and this in turn determines how we set about solving them. Our ability to see and understand a wicked problem is to an extent dependent upon our worldview and ethos.

Below: (p.252 –253)Figure 7. In this diagram, design process and examples of government policies which support them have been mapped along a continuum of leverage points which correspond to Meadows (2008). Based upon research done in collaboration with Chris Sherwin and Julie Richardson in 2005.

5. Information flows	4. Rules	3. Self-Organization	2. Goals	1. New paradigm
Communicating sustainability issues clearly, transparently	Involving relevant constituencies in design process	User needs/preferences inform socially inclusive design	Design that changes aspirations and social norms	Design for sustainable lifestyle informs the design of products/services
	• User-centered design • Co-design • Inclusive design	• User-centered design • Co-design • Inclusive design	• Service design (expanded)	• Industrial ecology (expanded) • Product service systems • Biomimicry (expanded)
Changing Consumption			**Changing Lifestyles**	
Product labeling schemes and standards		Participatory design of public services and infrastructure	Sustainable procurement directives	Design for sustainable lifestyle informs the design of products/services
Creating brand narratives and communications strategies that educate and promote sustainable choices	Raising awareness of and collaboration between the constituent groups that influence sustainable design and consumption	Design that changes aspirations and social norms	Developing narratives and educational messages that influence social norms and lifestyle aspirations	Helping to develop future scenarios for sustainable everyday life: future-casting

The importance of worldview

Meadows describes a system's most powerful leverage point for change as '...the shared idea in the minds of society, the great big unstated assumptions – unstated because unnecessary to state; everyone already knows them – constitute that society's paradigm or deepest set of beliefs about how the world works...' (Meadows 2008, pp.162–63). Fritjof Capra in his book *The Turning Point* says of society's most pressing problems:

> ultimately these problems must be seen as just different facets of one single crisis, which is largely a crisis of perception. It derives from the fact that most of us, and especially our large social institutions, subscribe to the concepts of an outdated worldview, a perception of reality inadequate for dealing with our overpopulated, globally interconnected world. (Capra 1982, pp.xvii–xviii)

The implication for designers is that a new way of seeing and thinking about problems is called for – a new design paradigm. Such a paradigm must be underpinned by a holistic worldview or ethos (Kearney 1984, Woodhouse 1996, Capra 1983, Clark 2002, Rozak 2001).

Since the scientific revolution of the seventeenth century, a reductionist and mechanistic style of thinking has predominated (Berman 1981, Capra 1996, Toulmin 1990, Rozak 2001) and its influence has impacted on virtually every discipline and profession, including design (Orr 1994, pp.104–11). This style of thinking leads designers to assume that the behaviour of a system of any kind can be understood – even predicted – in terms of the properties of its parts. Many design solutions are based upon predictions that are the result of relatively little research and understanding of both problem and context. Complex, living systems however, cannot be understood in these terms; their parts (which are often fragments of larger wicked problems) can only be understood within the context of the greater wholes of which they are part. As Capra emphasised in *The Web of Life:*

> ... systems thinking is 'contextual' thinking; and since explaining things in terms of their context means explaining them in terms of their environment, we can also say that all systems thinking is environmental thinking. (1996, p.37)

A logical follow-on to this might be 'all systems design should be environmental design.'

A new ethos in which the interdependencies and interconnections of the physical world are understood and honoured necessitates an entirely new way of designing predicated upon a fundamentally different concept of our world. Thomas Berry and Brian Swimme in their book *The Universe Story* (1992, p.243) have said *'the universe is a communion of subjects rather than a collection of objects'* which speaks of an ethos of interrelatedness, reciprocity and concern for 'other' that philosopher Martin Buber described as an 'I-Thou' relationship (2000). Understanding living systems principles and the anatomy and dynamics of wicked problems is a crucial step in designing sustainable solutions, but it isn't enough. Unless this understanding is underpinned by a holistic worldview any attempt to design a sustainable solution will fall short of its potential.

In Figure 7, biomimicry, (an approach which 'mimics' design solutions from nature. See Benyus 1997) has been mapped twice, at opposite ends of the spectrum. When applied from within a reductionist mindset, biomimicry can be used to improve the structure or basic configuration of a design, but within such a narrow application, it can be implicated in the exacerbation of wicked problems. Biomimicry has been used extensively by engineers as a design approach (Brebbia 2008), however the solutions developed are often implicated in larger wicked problems such as the development of toxic materials used in manufacturing processes, experimentation on/cruelty to animals and government surveillance. When biomimicry is applied within simplified contexts, designers do not have to take into account the larger ethical, social and environmental implications of their solutions. When biomimicry is applied from within a holistic worldview however, an organism is studied in context and its reciprocal relationships with its environment are seen as intrinsic to its 'design'.

Another example of the way in which a worldview can affect the design of solutions is healthcare. The allopathic, reductionist approach to healthcare views the body as a collection of parts or separate systems which can be treated in isolation by medical specialists (heart specialists, neurologists, gastrointerologists, and so on). Synthesised drugs are used to treat symptoms (the root cause of illness can go undiagnosed/untreated), which can have side effects as serious as the condition they have been prescribed to treat. The parallels to the application of dispersants in the gulf are obvious. A contrasting worldview or mindset is exemplified by alternative medicine, with acupuncture being one approach. The acupuncturist sees the body as an interconnected, interdependent system and looks for congestion or blockages that affect the whole organism. He sees symptoms not as problems to be treated, but rather as clues to understanding the root cause of the illness and which in turn suggest the type of treatment (design solution). The application of needles along 'meridians' can be likened to Meadows' leverage points within the system and the acupuncturist is always looking for the most powerful leverage point. Heretofore traditional design process can be said to have followed a reductionist model, but in order to address wicked problems and design sustainable solutions, it needs to take a more holistic, iterative approach.

A new design paradigm

One of the most fundamental changes for designers and design process will be a shift in focus from objects to relationships, which is the essence of systems thinking. This shift will require both a change in mindset as well as the acquisition of new knowledge (see Figure 8). An organic model of society and the environment will replace the dominant, mechanistic one and this in turn will suggest a more respectful, iterative and inclusive process for designing solutions. Ethics will be at the center of the design process and social and environmental concerns will be considered as important as profit to both designers and their clients.

The transition toward a new design paradigm has already begun. An increasing number of organisations are using design to develop sustainable solutions and professional design organisations are beginning to incorporate social and environmental concerns into their basic approach. Ezio Manzini was one of the pioneers of alternative design solutions and his ground-breaking projects Sustainable Everyday (2003), the Changing the Change Conference (2008) and the DESIS sustainable design network are examples of design for social innovation that involve transdisciplinary collaborative efforts that are based locally and connected globally. The O2 Global Network is a resource for designers wanting to work sustainably and both the Worldchanging and Design 21 websites offer valuable information about designers working for social change and the environment. Educational institutions such as Schumacher College in Devon, England and at Bija Vidyapeeth in the Doon Valley of India have for more than a decade offered education in a new worldview and ethos, and the Center for Ecoliteracy in Berkeley is a leader in the 'green schooling movement' which is educating the next generation of transition designers.

Design touches every part of our modern, everyday lives and is implicated in virtually every wicked problem confronting us today. Its sheer ubiquity speaks of its potential for positive change. Designers from all walks of life, working from within a new ethos and design paradigm can develop meaningful solutions for wicked problems and can create the initial conditions for sweeping positive change for society and the environment.

Understanding the interconnected, interdependent nature of reality
Placing an emphasis on cooperation and relationship
Acknowledging/respecting the intrinsic nature of all life forms
Developing the ability to think in long horizons of time
Understanding the principles of self-organisation and emergence
Understanding the power of limits
Acknowledgment that ignorance is part of the human condition
Acknowledgment of the limits and consequences of science and technology
Embracing a new educational model that acknowledges the challenges of 'transitional times' and is based upon co-learning and re-skilling

Figure 8. Characteristics of a new ethos. Sustainable solutions of any kind must arise out of a new mindset or worldview in order to realise their potential for change. Such a new mindset will be characterised by some of the characteristics listed here.

References

Archer, Bruce (1984) 'Systematic Method for Designers,' in *Developments in Design Methodology*, Nigel Cross, ed., John Wiley & Sons, Chichester.

Augros, Robert & George Stanciu (1987) *The New Biology*, New Science Library, Shambhala, Boston.

Benyus, Janine (1997) *Biomimicry*, William Morrow & Co., New York.

Berman, Morris (1981) *The Reenchantment of the World*, Cornell University Press, Ithaca.

Bija Vidyapeeth Earth University — at: http://

www.navdanya.org/earth-university.

Brebbia, C.A. (2008) ed., *Design and Nature,* WIT Press, Southampton.

Briggs, John & Peat, F. David (1989) *Turbulent Mirror,* Harper & Row Publishers, New York.

—, & Peat, F. David (1999) *Seven Life Lessons of Chaos,* HarperCollins, NY.

Buber, Martin (2004) *I and Thou,* Continuum, London.

Buchanan, Richard (1995) 'Wicked Problems in Design Thinking', in *The Idea of Design,* Victor Margolin & Richard Buchanan, eds., The MIT Press, Cambridge.

Capra, Fritjof (1982) *The Turning Point,* Flamingo Publishing, London.

—, (1996) *The Web of Life,* HarperCollins, London.

—, (2001) essay: 'Trying to Understand: A Systemic Analysis of International Terrorism', at: http://www.fritjofcapra.net/articles100501.html, October 5.

Changing the Change Conference (July 2008), Turin. At: http://emma.polimi.it/emma/showEvent.do?idEvent=23

Clark, Mary E. (2002) *In Search of Human Nature,* Routledge, New York.

—, & Sandra A. Wawrytko (1990) *Rethinking the Curriculum,* Greenwood Press, Westport.

Design 21 — Social Design Network, at: http://www.design21sdn.com/

DESIS Network — Design for Social Innovation and Sustainability, at: http://www.desis-network.org/

Fahrenthold, David A. & Steven Mufson (2010) 'Documents Indicate Heavy Use of Dispersants in Gulf Oil Spill,' *Washington Post,* August 1, Washington DC. At: http://www.washingtonpost.com/wp-dyn/content/article/2010/07/31/AR2010073102381.html,

Global Footprint Network (2010) *Ecological Footprint Atlas 2010,* at: http://www.footprintnetwork.org/en/index.php/GFN/page/ecological_footprint_atlas_2010, Global Footprint Network, Oakland.

Goodwin, Brian (1994) *How the Leopard Changed its Spots,* Charles Scribner's Sons, New York.

Gray, Rob & Markus Milne (2004) 'Toward Reporting on the Triple Bottom Line: Mirages, Methods and Myths,' in *The Triple Bottom Line, Does it All Add Up?* Adrian Henriques & Julie Richardson, eds., Earthscan, London.

Imhoff, Daniel (2005) *Paper or Plastic,* Sierra Club Books, San Francisco.

Jones, Christopher J. (1980) *Design Methods,* John Wiley & Sons, New York.

Kearney, Michael (1984) *World View,* Chandler & Sharp Publishers, Novato.

Korten, David C. (1999) *The Post Corporate World,* Berrett-Koehler Publishers, San Francisco.

Manzini, Ezio & Francoise Jegou (2003), *Sustainable Everyday: Scenarios of Urban Life,* Edizioni Ambiente srl, Milan. At: http://www.sustainable-everyday.net/SEPhome/home.html

Meadows, Donella (2008) *Thinking in Systems*, Chelsea Green Publishing, White River Junction.

Newsweek (2010) 'What the Spill Will Kill', at: http://www.newsweek.com/2010/06/06/what-the-spill-will-kill.html, June 6.

O2 Global Network, International Network for Sustainable Design, at: http://www.o2.org/index.php

Orr, David (1994) *Earth in Mind*, Island Press, Washington DC.

Papanek, Victor (1995) *The Green Imperative*, Thames and Hudson, London.

Protzen, Jean-Pierre & David J. Harris (2010) *The Universe of Design: Horst Rittel's Theories of Design and Planning*, Routledge, New York.

Rittel, Horst (1972) 'On the Planning Crisis: Systems Analysis of the "First and Second Generations",' *Bedrifts Okonomen*, no.8, Oct. 1972.

—, & M. Webber (1973) 'Dilemmas in a General Theory of Planning,' in *Policy Sciences*, Volume 4, Elsevier Scientific Publishing Company, Amsterdam.

Rozak, Theodore (2001) *The Voice of the Earth: An Exploration of Ecopsychology*, Phanes Press, Grand Rapids.

Salahi, Lara, Jane Allen & Seth Trueger (2010), 'Louisiana Locals Worry about Oil Spill's Health Effects', ABC World News, June 7, at: http://abcnews.go.com/Health/Asthma/gulf-oil-spill-potential-health-effects-people-louisiana/story?id=10849577.

Schumacher College — 'Transformative Learning for Sustainable Living,' at: http://www.schumachercollege.org.uk/

Simon, Herbert (1988) *The Sciences of the Artificial*, The MIT Press, Cambridge.

Swimme, Brian & Thomas Berry (1992) *The Universe Story*, Harper, San Francisco.

Toulmin, Stephen (1990) *Cosmopolis: The Hidden Agenda of Modernity*, University of Chicago Press, Chicago.

Wheatley, Margaret (2006) *Leadership and the New Science*, Berrett-Koehler Publishers, San Francisco.

Woodhouse, Mark B. (1996) *Paradigm Wars*, Frog Ltd., Berkeley, CA.

Worldchanging — at: http://www.worldchanging.com/

Yoon, Limi, Sanghyuk Yoon, Neo Martinez, Rich Williams & Jennifer Dunne (2005) 'Interactive 3D Visualisation of Highly Connected Ecological Networks on the WWW,' ACM Symposium on Applied Computing, March 13–17, Santa Fe. At: http://delivery.acm.org/10.1145/1070000/1066950/p1207-yoon.pdf?key1=1066950&key2=2143383921&coll=DL&dl=ACM&CFID=4551854&CFTOKEN=37231871

Zero Emissions Research and Initiatives, (2010) — at: http://www.zeri.org.

19. A Co-operative Inquiry into Deep Ecology

ESTHER MAUGHAN MCLACHLAN AND PETER REASON[1]

The Masters degree in Responsibility and Business Practice, offered at the University of Bath from 1997–2010 addressed the challenges facing society in integrating successful business practice with a concern for social, environmental and ethical issues.[2] It looked at the complex relationship between business decisions and their impact on local and world communities and economies, on the environment and on the workplace itself, helping participants develop management practices which are responsive to pressures for greater awareness in these areas.[3]

The course programme comprised eight intensive week-long workshops over two years. Each workshop delved deeply into a particular topic, with presentations, exercises and discussion. The workshop topics included: the globalisation of world economy and culture; the way our economic theories and policies place value on certain activities and ignore others; the capacity of corporations to act in sustainable ways and as corporate citizens; the development of meaningful work.

The staff team, when they originally designed the programme, was adamant that the programme, while clearly a business programme in a prestigious business school, should attend to questions of meaning, value, spirit, and in particular that students should be exposed to radical thinking about the nature of the planet Earth which is the originator of all human and non-human wealth. We wanted to explore deep ecology and Gaia theory and offer students an opportunity for a direct experience of the wildness of the natural world, as far as that is possible in the overcrowded British Isles.

To this end we teamed up with colleagues at Schumacher College in Devon, and in particular with the resident ecologist Stephan Harding.[4] Together we designed a week-long experience which included lectures on deep ecology (Devall & Sessions 1985; Macy & Brown 1998; Naess

1990), Gaia theory (Harding 2009; Lovelock 1979, 1988, 1991, 2006) and the state of the natural world, but where a lot of time was spent outside. We took participants on a night walk through woodland and spent an afternoon meditating by the River Dart. We summoned the Council of All Beings, the ceremony developed by John Seed and Joanna Macy (Macy & Brown 1998; Seed, Macy, Fleming, & Naess 1988) in which participants speak as the many diverse beings of their concern for the state of the world. And we spent one whole day in a hike along the upper reaches of the River Dart, along what must be one of the last remaining stretches of wilderness in England. On this walk we left the footpaths to scramble over rocks and under branches. We helped each other through bogs and over torrential streams. And under Stephan's guidance we experimented with deep ecology exercises: imagining how the world that we sense is also sensing us (Abram 1996); guiding each other in pairs on a blindfolded experience of the trees, rock, and mud; identifying with a being in the natural world and exploring through imaginative meditation how it is part of the cycles of Gaia (Reason 2007a, 2007b).

Co-operative inquiry

The whole MSc programme is designed using action research as a basis for learning, and throughout the programme there is an emphasis on inquiry processes and skills (Marshall, Coleman & Reason 2011). This deep ecology workshop is designed using the format of co-operative inquiry (Heron & Reason 2001; Reason 2003), which is a form of collaborative action research practice – research *with* rather than *on* people (Reason & Bradbury 2001, 2008). In traditional research, the roles of researcher and subject are mutually exclusive. The researcher only contributes the thinking that goes into the project, and the subjects only contribute the action to be studied in a relationship of unilateral control. In co-operative inquiry all those involved work together as co-researchers and as co-subjects.

Co-operative inquiry draws on an 'extended epistemology', seeing that human persons participate in and articulate our world in at least four interdependent ways: experiential, presentational, propositional and practical. *Experiential knowing* is through direct face-to-face encounter with a person, place or thing; it is knowing through empathy and resonance; *presentational knowing,* which grows out of experiential knowing, provides the first form of expression through story, drawing, sculpture, movement, dance, drawing on aesthetic imagery; *propositional knowing,* is 'knowledge about', expressed in concepts and ideas; and *practical knowing,* consummates the other forms of knowing in action in the world (Heron 1996; Heron & Reason 2008). The process of co-operative inquiry can be seen as cycling through four phases of reflection and action, in each of which a different way of knowing holds primacy.

In Phase 1 a group of co-researchers come together to explore an agreed area of human activity. In this first phase they agree on the focus of their inquiry, and develop together a set of questions or propositions they wish to explore. They agree to undertake some action, some practice, which will contribute to this exploration, and agree to a set of procedures by which they will observe and record their own and each other's experience. Phase 1 is primarily in the mode of propositional knowing.

In the deep ecology workshop the focus of inquiry was established as part of the course content. The questions posed for the week were: 'What is the experience of deep ecology?' and: 'What activities and disciplines aid its development?' Within these broad questions individual participants were invited to develop their own specific questions as the week progressed. The propositional knowledge on which the inquiry was based included the ideas about deep ecology and Gaia theory offered by Stephan.

In Phase 2 the co-researchers now also become co-subjects; they engage in the actions agreed and observe and record the process and outcomes of their own and each other's experience. In particular, they are careful to notice the subtleties of experience, to hold lightly the propositional frame from which they started so that they are able to notice how practice does and does not conform to their original ideas. This phase involves primarily practical knowledge: knowing how (and how not) to engage in appropriate action, to bracket off the starting idea, and to exercise relevant discrimination.

Starting with the night walk the evening we arrived at Schumacher College, participants were invited into the range of activities outlined above. As faculty we designed activities through which they could engage with the natural world in novel ways – to enter into relation with trees, to walk on the earth as a living being, to meditate with the river, to speak as a slug or as an oak tree ...

Phase 3 is in some ways the touchstone of the inquiry method. The co-subjects become fully immersed in and engaged with their experience. They may develop a degree of openness to what is going on so free of preconceptions that they see it in a new way. They may deepen into the experience so that superficial understandings are elaborated and developed. Or they may be led away from the original ideas and proposals into new fields, unpredicted action and creative insights. Phase 3 involves mainly experiential knowing, although it will be richer if new experience is expressed, when recorded, in creative presentational form through graphics, colour, sound, movement, drama, story, or poetry.

For many participants experiential knowing was the key to their learning. For many, living for a week in community in an area of amazing natural beauty, having time just to sit by a river, and being given permission to open themselves to the voice of the more-than-human world was of great significance.

In Phase 4, after an agreed period engaged in phases two and three, the co-researchers re-assemble to consider their original propositions and questions in the light of their experience. As a result they may modify, develop or reframe them; or reject them and pose new questions. They may choose, for the next cycle of action, to focus on the same or on different aspects of the overall inquiry. The group may also choose to amend or develop its inquiry procedures – forms of action, ways of gathering data – in the light of experience. Phase 4 is primarily the stage of propositional knowing, although presentational forms of knowing will form an important bridge with the experiential and practical phases.

The course participants were allocated to small groups (who also work together each day on simple household tasks to maintain the ecology of the College) which met at the end of each day to review and make their sense of the experiences. We encouraged participants to help each other articulate what has been important for them, to write reflectively, to draw or otherwise create visual images.

In a full inquiry the cycle will be repeated several times. Ideas and discoveries tentatively reached in early phases can be checked and developed; investigation of one aspect of the inquiry can be related to exploration of other parts; new skills can be acquired and monitored; experiential competencies realised; the group itself becomes more cohesive and self-critical, more skilled in its work. Ideally the inquiry is finished when the initial questions are fully answered in practice, when there is a new congruence between the four kinds of knowing. It is of course rare for a group to complete an inquiry so fully.

The deep ecology workshop was designed with three cycles of inquiry:

—discussion of the philosophy of deep ecology followed by an afternoon in meditation with the River Dart;
—an introduction to Gaia theory and the state of the world followed by the Council for All Beings;
—the day-long eco-walk down the River Dart with mini-talks and exercises.

Each of these cycles was followed by a review in small groups, and on the final morning we met as a whole group. Participants responded in turn to the questions: 'What is the experience of deep ecology? and: 'How do you get there?' The group as a whole discussed the responses and clustered them under headings. This reporting session was an energetic affair, full of both laughter and tears. This session was audio recorded and forms the basis of this article.

Our purpose as faculty in using the co-operative inquiry model was twofold: to continue our emphasis that the Masters course is based on a process of mutual inquiry in which all learn; and to formally introduce and teach the form of co-operative inquiry to make it available as an approach for course participants in their own work. On this workshop we as faculty used our authority to structure much of the inquiry process; it was not an inquiry in full collaborative form. We did this as a teaching and learning device, and because we believed it appropriate for us as faculty to use our authority and experience in the service of learning, while being open to feedback and comment from the group.

In what follows we draw on the audio recording made on the fourth MSc programme to show the range of learning experiences. While we have not worked systematically with the experiences of other groups our impression is that the learning is broadly representative of all twelve groups. MSc4 included ten men and fourteen women with ages between early 20s and late 50s, from nine nationalities living in seven countries from Finland to Vietnam. At that time they were working mainly in the corporate sector and in multi-nationals such as Rio Tinto and Barclays; a minority were independent consultancy and working for NGOs and local government.

The experience of deep ecology

The experience of deep ecology started for most of us with a true appreciation, as if for the first time, of the simple beauty of the more-than-human world versus the human-made urban world many of us live in. This experience is one of profound joy expressed by one participant as 'post human exuberance, when you sit on a rock and feel happy, it's not like when you're happy because you've had a birthday present, it's a different, more profound sort of happiness'.[5]

Beauty in this sense does not include the merely aesthetic. Deep ecology is 'awe at the proliferation and richness of living things' and at the more-than-human world's wondrous self-organisation: 'Everything finds its own place, one plant seems to be in just the right spot, there's no other place it could be'.

What is the experience of deep ecology?
• The experience of deep ecology is a feeling of joy and awe at the beauty of the more-than-human world.
• It is an appreciation of the delicate balance between chaos and order.
• It is the acknowledgment of the interconnectedness of all living beings, including ourselves, in the endless cycles of the planet.
• This acknowledgment leads to the direct identification of ourselves with other living beings and a redefinition of our place, no longer dominating nature but one equal part of it.
• It is a sense of the consciousness of other living beings and the reciprocal relationship between us.
• The experience is both of the moment and of eternity.
• The experience is that of a spiritual quest to reconnect with our true human nature and break down the artificial barriers we have erected.
• It is the feeling of home-coming.
• It is the celebration of the creator.

It was clear that for the city-based participants particularly, an obvious first step towards this experience was simply to 'take time to be in the non-human world'. We began to question why such a large part of our lives is spent indoors in contrast to the week at Schumacher College when on average half a day was spent outdoors. However, it was agreed that being in the more-than-human world was not enough unless we were truly open to experiencing its magic, not just through sight but with all our senses: 'Deep ecology is about using my senses and my intuition to actually connect with what is happening with the rhythms of life, it's being still and touching the wonder'. It was only through this intensity of focus, which felt difficult for some but easy for others, that we were able to cut through the weight of our preoccupations and preconceptions and experience the more-than-human world in a way that bypassed our reason and made connections at a deeper emotional level.

We found beauty in 'the wonder and magic of nature's complex cycles'. Through cycles of birth, death and re-use we became aware that 'everything is related in one way or another' and deep ecology provides us with an 'understanding of the intimate relationships which exist and which we have with nature as well'. Our 'connectedness to the rhythms of the natural world' is something which our urban lives allow us to forget and the experience of deep ecology places us back within our most fundamental context: 'we are nature'. One participant elaborated on this: 'I thought the core experience was to actually feel myself as part of the natural world. I don't think we normally actually feel that'.

This interconnectedness created in some a sense of perfect balance, so often missing in our own personal lives: 'Deep ecology is the opposite of the unstable equilibrium that we try to live with. We've fallen over, we need to get back to the balance that we once had, where we could live life in a much richer and fuller sense'. This idea of balance is inherent in natural cycles where nothing is ever wasted and one participant gave us the flippant yet sobering reminder that 'we are recycled and should make the effort to treat our bodies well and so become good compost'. As another example of this balance, for others deep ecology was 'the acknowledgment of order in chaos and chaos in order', 'when you can allow your conscious and subconscious mind to become aware of the controlled chaos of natural systems'. This has parallels with the deep ecology experience itself which was for many of us a turbulent one: 'I feel like I've been sitting on the edge of chaos all week'.[6]

It seemed that this aspect of deep ecology was one that was particularly fostered by more formal scientific learning, that is, propositional knowing. Stephan's lectures were original and exciting and his passion for cycles inspired us to see them at work for ourselves outside the classroom. For some of us, propositional knowing is vital on the path towards the experience of deep ecology, if only because of its familiarity from traditional education. It was important to complement our experiential knowing with a more conceptual framework and one participant spoke of her 'relief and excitement that there are now facts I can share with others'.

Another claimed: 'I don't like Maths particularly, I hated statistics in my first degree but I found through Stephan that I could actually see some beauty in mathematics'. The importance of wise teachers and elders in the journey towards the deep ecology experience cannot be overemphasised; their inspiration and guidance was crucial to the success of the week.

A greater understanding, both rational and intuitive, of the interconnectedness of nature's cycles leads us to re-evaluate our own roles within those cycles. The afternoon spent in quiet beside the River Dart highlighted this for one member of the group: 'I really got a sense of the busyness of what's going on at a not-human level. I had never before appreciated the rich detail of life's activity happening without any reference to us humans'. The experience of deep ecology is therefore 'to redefine what it means to be human, we are not dominant'. Others experienced deep ecology in a similar way, as a 'knowing of nature's secrets, they've been unlocked for me and now I know that I am part of this experience and this is my story too'.

One participant expressed a commonly held view: 'Before I came on this week I had my doubts because I always felt that I had an affinity with nature but was outside nature, not necessarily a part of it. Now I see we're all equal parts of the same earth, there's the interconnectedness of one big family'. Another participant agreed: 'The essence of deep ecology is seeing yourself as a part not as an observer, and so moving from knowing truths to feeling truths. It is still seeing but it's also smelling, touching, feeling and sensing, getting the whole of yourself into it. What was really powerful for me was putting myself in the place of another being and looking at myself in the mirror'. Deep ecology is therefore the experience of personally 'relating to everything'. This transition from unengaged observer to engaged participant is paralleled in the approach of action research (see above) as is the acknowledgment that there is more than one way of knowing truths.

Deep ecology is about realising 'the dependency of all beings upon each other' and that 'every living thing has a purpose'. This leads to the questioning of one's personal sense of purpose and ultimately to a redefinition of one's personal identity. We discovered that 'the experience of deep ecology is about the spiritual quest to really reconnect with our true human nature' and this was expressed in other ways: 'A journey of the self and a journey to the self', 'Realising I have a role to play' and 'Finding peace and inner self'. This led to a growing understanding that our human viewpoint is just one of many equally valuable perspectives and it provided us with the realisation that the more-than-human world is responding and reaching out to us in turn. We had complementary experiences, at once both a melting of the barriers between ourselves and other non-human beings and a heightened sense of the conscious, separate life of those beings.

We found our experience was particularly heightened by the exercises during our day-long wilderness walk when we were invited to close our eyes, touch our surroundings and sense our surroundings touching and feeling us in reply. One participant spoke of 'the blur between me and the moss I was touching, it was difficult to know where I ended and the moss began. Then there was the exercise where we really probed our surroundings, I almost felt like asking permission of this other living entity, 'May I?' and 'Should I?' and 'I've never done this before'. 'I really experienced a wonderful balance between the blur and the sense of otherness, in our existence, our relationships with the living world, our very being'. This notion of otherness was also expressed in this way: 'Now I know the earth and everything on it has a heart and has feeling'.

Throughout the week we felt welcomed by the more-than-human world and many of us shared this participant's feeling 'of coming home, of being accepted by the place like when I've had a really happy home, I've just walked in and been embraced'. One participant described the experience of deep ecology as 'a mutual "letting-in-ness" where nature lets you in on all its huge libraries of knowledge and you are willing to be let in'. Another participant admitted: 'I've now recognised that I've got everything I've ever asked for whenever I've gone to my special place to try and work things out. I might not have realised it at the time, but it's all been there for me to take'.

We spoke of our surprise and pleasure that the more-than-human world was soft and sensual rather than painful and frightening as we are sometimes brought up to believe. But it was only through 'active, strenuous physical engagement' with the more-than-human world that these experiences were made possible, 'by fitting yourself into the nooks and crannies'. One group member spoke of the artificial barriers we erect between ourselves and the more-than-human world when she said: 'Walking along the river, particularly the clambering, reminded me how neat and tidy we are invited to keep ourselves and how we never exert ourselves or get dirty. There's all sort of things we don't do, which stop us making the connection to the bigger picture'.

This breaking down of both physical and emotional barriers,

'allowing the armour or uniform to fall away', is a vital component for the achievement of the deep ecology experience and a developing sense of the reciprocal relationship between ourselves and the more-than-human world led some of us to begin a new kind of dialogue. One group member spoke emotionally of his personal route to the deep ecology experience: 'Take your poor battered heart into the wilderness and when you're there, you listen to the wind and you ask and you listen and you ask and you listen, and that's how you get there'.

Throughout the week we explored the Gaian concept of the living world as a single conscious entity able to express emotions such as happiness. Whether we chose to interpret this literally or metaphorically we shared this participant's view: 'Something that's been moving for me is that, having talked about qualities, I really got the notion of a happy wood, I really understood it when Stephan pointed out that it was so diverse and full of life and abundant and growing. I think all the billions we spend on tourism and holidays show how we are yearning for this kind of happiness but actually we destroy it at the moment in our yearning'. One group member brought together the ideas of a conscious world and a reappraisal of personal identity in 'the notion of the ecological self, we humans are one of the parts of the universe which is conscious of itself so we are the universe looking at itself'.

One theme that is raised throughout the MSc course is that of timescales, from the short-termism of shareholders' expectations to long term visions of a new society, and the week at Schumacher prompted further consideration of this. Deep ecology is about 'experiencing the moment fully and being deeply connected to what's going on'. This notion of being fully present in and engaged with the moment is also one which is central to action research and is a practice many of us find hard during the self-imposed rush of our lifestyles. The way to achieve this experience of deep ecology is therefore to 'be still, be silent and appreciate the moment for its intrinsic value'. One group member elaborated this quality of the deep ecology experience: 'I experienced the moment rather than thinking "What have I been doing?" and "What am I going to do?" I was aware that I was just deeply connected with what was going on. But somehow, in that moment, I could perceive the past and future, I could see the past in that I could see where the rocks had come from, I could see the future in the sense that I could see where the river was going, and for me time did seem to stop'.

On the other hand, deep ecology provides 'a sense of eternity which is a big issue for a lot of people, they are concerned about how their memory will survive and whether they will leave a mark'. Deep ecology 'is timeless so it's the past, the present and the future and we need to understand all of them and explore the future in order to take ourselves there'.

The eternal quality and beauty of interconnected cycles and individual living beings led many of us to explore the spiritual nature of the deep ecology experience. One participant explained how the week at Schumacher clarified her beliefs: 'Years ago, when people used to say to me "Are you religious?" or "What's your faith?" I never had one. I was never able to say anything except: "Well, the only sense I've got is that I believe in nature", and that was when I started to realise that spirit is all around me. Deep ecology for me is about understanding nature, about understanding the bigger picture, it's about the spirit that's all around me, and it's in everything and everyone'. For those of us who are Christians, deep ecology provided 'A language and a means of meeting and celebrating the creator of all things'.

It also provides individuals with a sense of purpose in a spiritual context, as described by one participant who assumed the nature of a kestrel during the Council of All Beings: 'I began to think about my kestrel and I was beginning to wonder what the purpose of a kestrel is and yes, it controls small animals so they don't overrun, but in a way the purpose of a kestrel is to sense freedom, but more than that, to enjoy the sense of freedom. And I think that as you get more sentient you have a greater purpose to enjoy the senses you have and you enjoy those senses on behalf of creation, or creator, and for me personally that is God'.

Taking deep ecology into our lives

Inevitably many other questions arose in connection with our new experience but the most pressing was: 'How do you take it away and keep it for yourself?' Most of the group agreed that the deep ecology experience 'takes effort to allow into our lives and requires mental and spiritual preparedness. We're not all going to have transformational experiences in which we suddenly form a connection with the non-human world which instantly and permanently transforms our way of thinking'. Another participant acknowledged the difficulties: 'We're so distanced from nature's cycles, some of us are distanced from them scientifically as we don't know the facts and some of us are distanced from them because we just don't spend enough time outside'.

What activities and disciplines aid the development of deep ecology?
• Spending time outside, preferably in the wilderness, in a state of openness.
• Physically engaging with other living beings which requires us to abandon our cultural preconceptions and overcome negative emotions such as embarrassment and cynicism.
• The guidance of inspirational teachers and wise elders.
• Being alone and/or having the support of a likeminded group, it varies for different individuals.
• Personal practices such as meditation and free-fall writing.
• A combination of the many ways of knowing, both emotional and rational, and an acknowledgment that different individuals will take different paths to the experience.
• Ongoing effort and commitment to integrate the deep ecology experience into our lives and be aware of our responsibilities to the more-than-human world.

One proposed solution to these difficulties was the continuing of practices such as creative thinking, meditation and free-fall writing which some of us were introduced to during the week: 'There has to be a yearning for this approach to become part of our skill, part of our practical knowledge. I think that's very important because we will be making our own cycles in our lives and we must introduce interconnectedness of thinking there, too'.

Another difficulty to overcome was that of self-consciousness and concern at the judgments of others, something which many of us were acutely aware of at the beginning of the week. One participant whose discovery of his ecological self was particularly emotional admitted: 'People were saying to me last Friday when I was going on this course, that's the tree-hugging part of the course, you'll being flapping around and free-associating and I thought: "Absolutely no chance, no way will I be doing any tree-hugging". How wrong was I!' However, by the end of the week we all felt much more at ease with this issue as one group member admitted: 'I don't really care now if people think I'm completely mad, sitting in the middle of a field on my own, reflecting, in silence'.

One thing that we all agreed upon was the importance of striving to enable others to experience a similar transformation: 'I feel very strongly that deep ecology is not about a club, it shouldn't be a secret or the privilege of a few, it's about rolling it out, making it accessible, so it becomes an experience for the many'. For some the experience of deep ecology was essentially about 'getting engaged to a movement to create new thinking and to provide a new way of providing solutions to a society in search of trust'. The transformative power of simply sharing the deep ecology experience with other members of the group was also noted: 'I really enjoyed sharing everything with the group; this was what helped me to feel a lot better about myself'. Another group member highlighted how the blindfold exercise fostered trust at an individual level; for another 'the experience of how to get there is, for me, through overcoming loneliness and through fellowship'.

Ultimately to maintain the deep ecology experience requires 'a commitment to the choice to live in a certain way' and we agreed that 'deep ecology is not a skin deep thing, it's not about putting on a pair of boots and walking outside, it's really a deep change, a deep commitment throughout one's life, based on fact and a sense of spiritual awakening'. This deep change involves 'acting on new-found responsibilities' so that 'we begin to treat the earth as we would treat ourselves'. One participant expressed it like this: 'For me the week has first of all been about realising my place in all this, which I'm not sure I did before, and then along with realising my place, realising my responsibilities for being in that place'. This may take the form of a reappraisal of what is truly important and valuable in our lives and of 'living gracefully': 'Deep ecology is about remembering our vital needs and finding freedom those in vital needs'. When combined with a shared experience of deep ecology, this could lead to a 'collective experience of responsibility for the many'.

The experience of deep ecology was an emotional one for all of us, ranging from joy to confusion to frustration to sorrow. As our busy week demonstrated, there is no one path to the experience of deep ecology, and practices and exercises which work for one individual may leave another cold. This is why the planning of the co-operative inquiry and an acknowledgment of the four ways of knowing were so important; as one participant said: 'I felt I was being invited to approach this in any way I liked, and there was no demand for me to feel this or believe that, either spiritually or rationally. It was more about finding your own way'. The fact that there are so many ways to achieve the experience of deep ecology gave us all a sense of optimism which is sometimes missing on this course, dealing as it does with the catastrophes we have brought about because we falsely place ourselves outside the more-than-human world. 'I believe that deep ecology is an incubator for a new value system. Being able to reach people through both heart and intellect gave me a sense of hope. There's potential with this approach, we could actually reach anybody; how can someone say no to this as long as we're not pushing it and saying you must feel it in this way and not that? I don't know if anybody can turn it away'. It is important to the future we share with all living things that they don't.

Notes

1. This is a lightly edited version of an article published in 2001 (Maughan & Reason 2001). It is based on co-operative inquiry involving all members of this course. The conclusions of the inquiry were prepared for publication by Esther Maughan, a member of the group, who undertook the work of sorting the tape transcript into categories, selecting quotes and composing the text; and Peter Reason, who wrote the introductory material. The original article was agreed by members of the course, whose contribution we acknowledge with thanks.
2. A programme based on similar principles, the MSc in Sustainability and Responsibility, is at the time of writing offered at Ashridge Management College, UK.
3. For a description of the programme and narratives of the activities of graduates see Marshall, J., Coleman, G., & Reason, P. (2011) *Leadership for Sustainability: An action research approach,* Sheffield: Greenleaf.
4. Schumacher College is an international centre for ecological studies offering a range of educational opportunities including short courses and Master programmes in Holistic Science and Transition Economics. See www.gn.apc.org/schumachercollege/
5. All words in quotation marks are taken from the audiotape of the group's reflection at the end of the workshop.
6. The language on chaos and complexity was influenced by a talk by the late Brian Goodwin (Goodwin 1994, 1999).

References

Abram, D. (1996) *The Spell of the Sensuous: Perception And Language In A More Than Human World,* Pantheon, NY.

Devall, B., & Sessions, G. (1985) *Deep Ecology: Living As If*

Nature Mattered, Gibbs M. Smith, Salt Lake City.

Goodwin, B.C. (1994) *How the Leopard Changed its Spots: The Evolution Of Complexity,* Weidenfeld and Nicholson, London.

—, (1999) 'From Control to Participation via a Science of Qualities.' *ReVision,* 21(4), pp. 26–35.

Harding, S.P. (2009) *Animate Earth,* Green Books, Foxhole, Dartington.

Heron, J. (1996) *Co-operative Inquiry: Research Into The Human Condition.* Sage Publications, London.

—, & Reason, P. (2001) 'The Practice of Co-operative Inquiry: Research with rather than on people.' In P. Reason & H. Bradbury (eds.), *Handbook of Action Research: Participative Inquiry And Practice* (pp. 179–88). Sage Publications, London.

—, & Reason, P. (2008) 'Extending Epistemology with Co-operative Inquiry.' In P. Reason & H. Bradbury (eds.), *Sage Handbook of Action Research: Participative Inquiry And Practice.* Sage Publications, London.

Lovelock, J.E. (1979) *Gaia: A New Look At Life On Earth,* Oxford University Press.

—, (1988) *The Ages of Gaia: A Biography Of Our Living Earth,* W.W. Norton, NY.

—, (1991) *Gaia: The Practical Science Of Planetary Medicine,* Gaia Books, UK.

—, (2006) *The Revenge of Gaia,* Allen Lane, London.

Macy, J.R., & Brown, M.Y. (1998) *Coming Back to Life: Practices to reconnect our lives, our world,* New Society Publishers, Gabriola Island.

Marshall, J., Coleman, G., & Reason, P. (2011) *Leadership for Sustainability: An Action Research Approach,* Greenleaf, Sheffield.

Maughan, E., & Reason, P. (2001) 'A Co-operative Inquiry into Deep Ecology.' *ReVision,* 23(4), 18–24.

Naess, A. (1990) *Ecology Community and Lifestyle: Outline of an Ecosophy,* trans. D. Rotherberg, Cambridge University Press.

Reason, P. (2003) 'Doing Co-operative Inquiry.' In J. Smith (ed.), *Qualitative Psychology: A Practical Guide to Methods,* Sage Publications, London.

—, (2007a) 'Education for Ecology: Science, aesthetics, spirit and ceremony.' *Management Learning,* 38(1), pp.27-44.

—, (2007b) 'Wilderness experience in education for ecology.' In M. Reynolds & R. Vince (eds.), *The Handbook of Experiential Learning and Management Education* (pp.187–201) Oxford University Press.

—, & Bradbury, H. (eds.) (2001) *Sage Handbook of Action Research: Participative Inquiry and Practice,* Sage Publications, London.

—, & Bradbury, H. (eds.) (2008) *Sage Handbook of Action Research: Participative Inquiry and Practice* (2nd ed.), Sage Publications, London.

Seed, J., Macy, J.R., Fleming, P., & Naess, A. (1988) *Thinking Like a Mountain,* Heretic Books, London.

Index

Nature's Due
Healing Our Fragmented Culture

Brian Goodwin

'In understanding nature more deeply, we understand ourselves more profoundly. This book is a brilliant articulation of this process, pointing to the emergence of a new culture of co-operation and harmony.'
– David Lorimer, *Scientific and Medical Network Review*

'[Goodwin's book] has a breath-taking range of scholarship that takes the reader on a journey of discovery through cultural history, scientific history, paradigm change, modern systems theory, chaos theory, evolutionary biology and a new field called biological hermeneutics.'
– *Resurgence*

Nature's Due: Healing Our Fragmented Culture challenges modern ideas on the interaction of science, nature and human culture.

Brian Goodwin, acclaimed author of *How the Leopard Changed Its Spots*, proposes that, in order for us to once again work with nature to achieve true sustainability on our planet, we need to adopt a new science, new art, new design, new economics and new patterns of responsibility.

A New Renaissance

Transforming Science, Spirit and Society

David Lorimer

This book diagnoses an urgent need for change and renewal in a period of crisis for philosophy, science and society. It is apparent to many that reductionist science with its materialist values is losing credibility. Its objectives of growth and acquisition, and its guiding principles asserting that there is no intrinsic meaning to life or purpose in the cosmos, are now widely seen as creating an unsustainable world.

Contributions in the first part of the book diagnose the sources of the crisis in today's world. The second section searches for a new understanding of consciousness and mind, based on findings in recent non-materialist philosophy. The third section looks to a renewal of spirituality beyond religion, aiming to recapture the personal depth and connection to the cosmos that materialism denies or ignores. The fourth section examines possible reforms in politics, economics and education to help bring forth a society that can sustain the flourishing of human beings in the globally interconnected world of the twenty-first century.

The Global Brain

The Awakening Earth in a New Century

Peter Russell

'It is important that we gain a stronger sense of being able to choose our future rather than stumble into it. We have been set an evolutionary exam and have here a textbook which we as students can study.'
– David Lorimer, *Scientific and Medical Network Review*

'It ranks with the best of Fritjof Capra and Lyall Watson as part of a trend in communicating ideas about mankind's inner self and our relation to the physical world; it is very well written, and it deserves to be a best seller.'
– *New Scientist*

We've seen the power of the internet to connect people around the world in ways never before known. This remarkable book argues that the billions of messages and pieces of information flying back and forth are linking the minds of humanity together into a single, global brain: a brain with astonishing potential for the Earth.

Peter Russell weaves together modern technology and ancient mysticism to present a startling vision of the world to come, where humanity is a fully conscious superorganism in an awakening universe.

Sky and Psyche

The Relationship between Cosmos and Consciousness

Nicholas Campion

The relationship between the human soul and the stars has been central to the spiritual and esoteric traditions of Western thought, and many other cultures, for thousands of years. Medieval Christians thought that heaven was located above the earth, beyond the stars. Our modern society, however, has largely severed the relationship between the human spirit and the sky.

This book explores ideas, beliefs and practices which meet at the boundary of psychology and cosmology, the universe and human imagination. This book addresses this special relationship from a variety of challenging and inspiring approaches.

The contributors include James Hillman; Liz Greene; Professor Neville; Nicholas Pearson; Professor Jarita Holbrook; Dr Angela Vos; Bernadette Brady; Jules Cashford; Noel Cobb; Cherry Gilchrist; Robert Hand; and Professor Richard Tarnas.